METEORITES AND THE ORIGIN OF PLANETS

Earth and Planetary Science Series
Patrick M. Hurley, *Consulting Editor*

Ahrens: Distribution of the Elements in our Planet

Faul: Ages of Rocks, Planets, and Stars

Morisawa: Streams: Their Dynamics and Morphology

Wood: Meteorites and the Origin of Planets

METEORITES AND THE ORIGIN OF PLANETS

John A. Wood
Smithsonian Institution
Astrophysical Observatory

McGRAW-HILL BOOK COMPANY
New York
St. Louis
San Francisco
Toronto
London
Sydney

Meteorites and the Origin of Planets

Library of Congress Catalog Card Number 68-13886

1234567890VBVB7543210698

PREFACE

In the beginning there was nothing. There arose a swelling, a ferment, a black fire, a spinning of vortices, a bubbling, a swallowing—there arose a whole series of pairs of props, posts, or piles, large and small, long and short, crooked and bent, decayed and rotten. Similarly there arose pairs of roots, large and small, long and short, and so forth: there arose countless and infinitely many supports. Above all, there now arose the ground, the foundation, the hard rock, there arose the space for light, there arose rocks of different sorts.

—Marquesas Islanders' myth of creation *

Men have always wondered about the beginning of things. It is especially fascinating to try to stretch our comprehension back through billions of years to a time when the darkness and cold of space were driven back by our sun as it began to burn—to a time when the Earth we live on was born. Every society since the ancient Sumerians has believed that it had some understanding of the creation of the world. It might seem that we are not really in any better position to account for the origin of the Earth now than were the Sumerians or the Marquesans. How can we hope to know what happened so long ago? Surely all evidence of the first events has been swallowed up in the chasm of time that separates us from them, so that we, too, can only speculate.

Fortunately this is not quite the case. The theories of the origin of the solar system that are current are based on numerous scientific clues, bits of information provided prin-

* From K. von den Steinen, Reise nach den Marquesas Inseln, *Verhandl. Ges. Erdkunde zu Berlin,* vol. 25, pp. 489–513, 1898.

cipally by astronomy and astrophysics and by *meteorites.* Meteorites are pertinent to the question because they are demonstrably the most ancient, the most primitive pieces of planetary matter that we have access to at the present time—a billion years older than the oldest known Earth rock. In fact there is evidence, to be developed later in this book, that some of the meteorites have survived virtually unchanged from the time when the planets were born.

The significance of this fact probably escapes nongeologist readers. It is hard for a person not educated in mineralogy and geochemistry to conceive how much information can be contained in a single rock small enough to fit in the palm of his hand—information about the processes that formed the rock and have affected it since. In the case of the unchanged meteorites just mentioned, the processes that formed the rock are the processes that formed the planets, or at least one of them. Clearly it is of very great interest to us to read the record in the meteorites.

This is why we study the meteorites: not primarily because they come to us from outer space, but rather because of their ancientness. They have been studied intensively during the past decade, and a very substantial body of knowledge has accumulated. Unfortunately no universally agreed-upon picture of the origin of meteorites (and planets) has emerged from these data. The facts can be interpreted in terms of several radically different rock histories. Of course from the point of view of any particular student, including the writer, one particular interpretation usually seems immensely more plausible than the others. For this reason the reader should be forewarned that, although areas of controversy and alternative interpretations will be identified, the latter may not always get equal time in the following chapters.

John A. Wood

CONTENTS

1 STONES FROM THE SKY

> Several perſons at Wold Cottage, in Yorkſhire, Dec. 13, 1795, heard various noiſes in the air, like piſtols, or diſtant guns at ſea, felt two diſtinct concuſſions of the earth, and heard a hiſſing noiſe paſſing through the air; and a labouring man plainly ſaw (as we are told) that ſomething was ſo paſſing, and beheld a ſtone, as it ſeemed at laſt, (about 10 yards, or 30 feet, diſtant from the ground), deſcending, and ſtriking into the ground, which flew up all about him, and, in falling, ſparks of fire ſeemed to fly from it. Afterwards he went to the place, in common with others who had witneſſed part of the phænomenon, and dug the ſtone up from the place where it was buried about 21 inches deep. It ſmelled, as is ſaid, very ſtrongly of ſulphur when it was dug up, and was even warm, and ſmoked. It was ſaid to be 30 inches in length, and 28½ in breadth, and it weighed 56lb.

Edward King (1796)*

So far as we know, meteorites have always pelted the Earth. Philosophers of ancient Greece (Anaxagoras, Diogenes of Apollonia) and writers of the Han Dynasty described meteor-

* From Remarks Concerning Stones Said to Have Fallen from the Clouds, Both in These Days and in Ancient Times, *Gentleman's Mag.*, vol. 66, pt. 2, pp. 844–849.

ite falls. The sacred black stone of Kaaba, in Mecca, to which pilgrim Moslems pay homage, is apparently a meteorite. Radioactive dating techniques have shown that a few meteorite "finds" fell to Earth over 10,000 years ago.

Anaxagoras and other writers of classical times supposed meteorites to be heavenly bodies that somehow had been loosened from their moorings and then fell to Earth. During most of the centuries since their time, however, this view has not been widely held. Scholars either disbelieved altogether that stones could fall from the sky (the witnesses were, after all, usually unlettered peasants) or contrived one way or another to explain them as a terrestrial phenomenon:

> MAY it not be a reaſonable conjecture, that all the various ſubſtances which have fallen from the atmoſphere, in latter as well as in former times, are nothing more than the ſands, and other contents, found at the bottom of lakes and large rivers, and from the ſhores of the ſea, *naturally* produced by the powerful influence or the attraction of the clouds? It is but a trite obſervation to ſay, that the clouds make frequent viſits to the waters of the earth, from which they uſually carry away large quantities of that element, and with it, no doubt, the ſubſtances (even with ſome of the fiſh) which form the beds, in proportion to the heat of the weather, and the depth of thoſe waters which the clouds, when they fall, happen to attach upon. It is as ſelf-evident, that the ſtreams which aſcend with the clouds are ſometimes clear as cryſtal, at other times thick and muddy. When the latter is the caſe, then it is that theſe ſubſtances may be concreted; and, by ſome extraordinary concuſſion in the atmoſphere, return to the earth.

Near the end of the eighteenth century several meteorite falls were widely publicized, including the Wold Cottage fall (above), and these seem to have caused a surge of interest in "stones from the clouds." A number of writers, of whom Mr. Bingley is a fair example, offered explanations of the phenomenon. One author was possessed of substantially clearer vision than the others, however. This was E. F. F. Chladni, a German lawyer and physicist (acoustics). Chladni wrote a small book, which appeared in 1794, entitled *Observations on a Mass of Iron found in Siberia by Professor Pallas, and on other Masses of the like Kind, with some Conjectures respecting their Connection with certain natural Phenomena.* In it he argued that these iron masses are spent fireballs (in one instance the iron mass had been found immediately after a bright fireball was seen in the atmosphere), and that they were extraterrestrial in origin, wandering bits of interplanetary matter—possibly fragments of a broken planet—which had fallen to Earth. He correctly surmised that after such an object plunged into the atmosphere, air friction would heat it so intensely that it would glow brilliantly and melt, producing the fireball phenomenon. In a later paper (1799) Chladni extended these concepts to include stony meteorites such as the one that fell at Wold Cottage.

Chladni supposed that atmospheric friction melted the meteorites completely and was responsible for the markedly igneous character of the masses he had examined. However, now we know that this is not the case: meteorites are heated so briefly in the atmosphere that little warmth has a chance to penetrate to their interiors. Consequently they are never "red hot" when recovered, and only the outermost centimeter or less shows any evidence of thermal damage. Their surfaces do melt during atmospheric flight, but melted material is ablated off as fast as it forms and is left as a trail of droplets (Fig. 1-1).

W. Bingley (1796)*

* From Stones Fallen from the Air a Natural Phenomenon, *Gentleman's Mag.*, vol. 66, pt. 2, pp. 726–728.

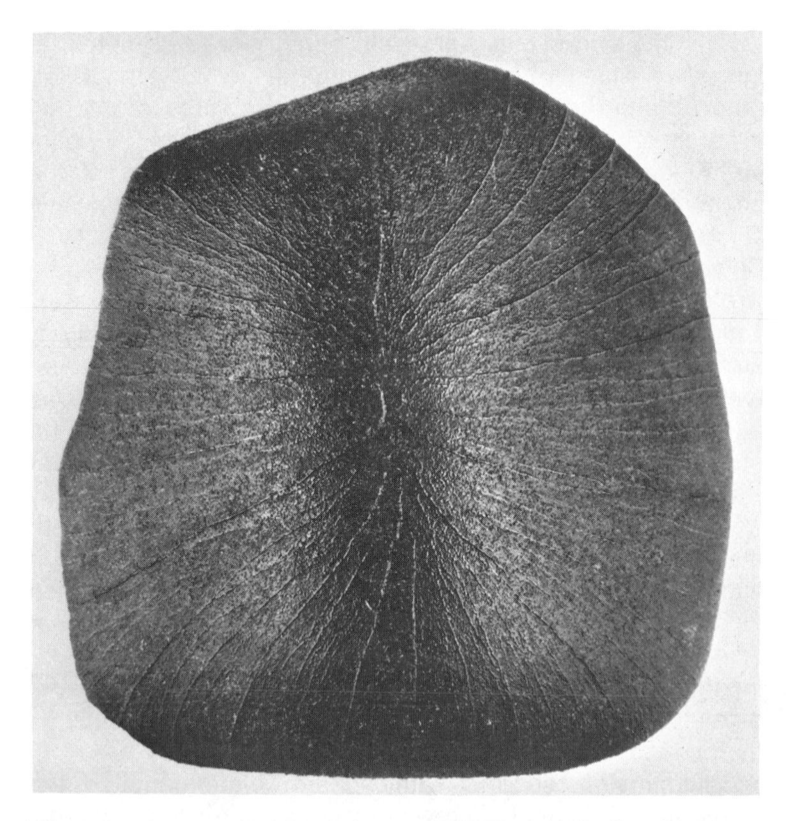

Fig. 1-1. An overall view of the small La Fayette (Indiana) meteorite, showing well-developed flight form and fusion crust. Streaks of melted rock that flowed back from the leading edge (center of photograph) can still be seen. Diameter, ~10 cm. (*Figure courtesy of Field Museum of Natural History.*)

Chladni's idea of extraterrestrial meteorites was a radical one, and he hesitated at first to expose himself to ridicule by publishing it. Yet the times seem to have been ripe for it, and within a decade it had won general acceptance. (In Europe, at any rate. Thomas Jefferson seems to have been a holdout: upon hearing of the Weston, Connecticut, meteorite fall, reported by Yale Professors Silliman and Kingsley in 1808, he is said to have remarked, "It is easier to believe that Yankee professors would lie, than that stones would fall from heaven.")

PRETERRESTRIAL ORBITS OF METEORITES

Only in relatively recent years has anything been learned about the paths in space followed by meteorites before they encountered the Earth. If the trajectory and velocity of a fireball in the atmosphere are known with sufficient accuracy, its path can be projected back into space and its former orbit determined. At first all that could be done in this direction was to interview eyewitnesses of a bright fireball, as many of them and as widely dispersed as possible, asking each to point out the course of the fireball as it appeared to him and to estimate the duration of its burn. Careful evaluation and collation of numerous such reports permitted a trajectory and orbit to be estimated. Obviously much uncertainty is attached to eyewitness reports by lay observers, and the orbits obtained are of little value. Three of the best-determined visual orbits for meteorites that were actually recovered are plotted in Fig. 1-2.

In 1959, by great good fortune, a fireball (meteorite) passed in front of multiple-station meteor cameras in Czechoslovakia, then fell at the town of Příbram, near Prague. The cameras were engaged in a systematic survey of very faint meteors in the upper atmosphere, but of course they also recorded the passage of the Příbram meteorite, and from the negatives it was possible to calculate with great accuracy the stone's preterrestrial orbit (Fig. 1-2). Příbram remains to this day the only instance where a precise orbit was determined *and* the meteorite recovered.

Additional information about orbits has come from photographic studies of meteors. Meteors are the dim "shooting stars" we see high in the atmosphere whenever small bits of interplanetary matter (*meteoroids*) encounter the Earth. Those studied photographically are commonly due to objects 0.1 to 1 g in mass. The meteoroids are not simply small meteorites: the great majority are very fragile, low-density objects (~0.25 g/cm^3 versus ~3.6 for stony meteorites), and are apparently bits of cometary debris. Occasionally high-density meteoroids are recorded, however (the difference in density can be deduced from their behavior in the atmos-

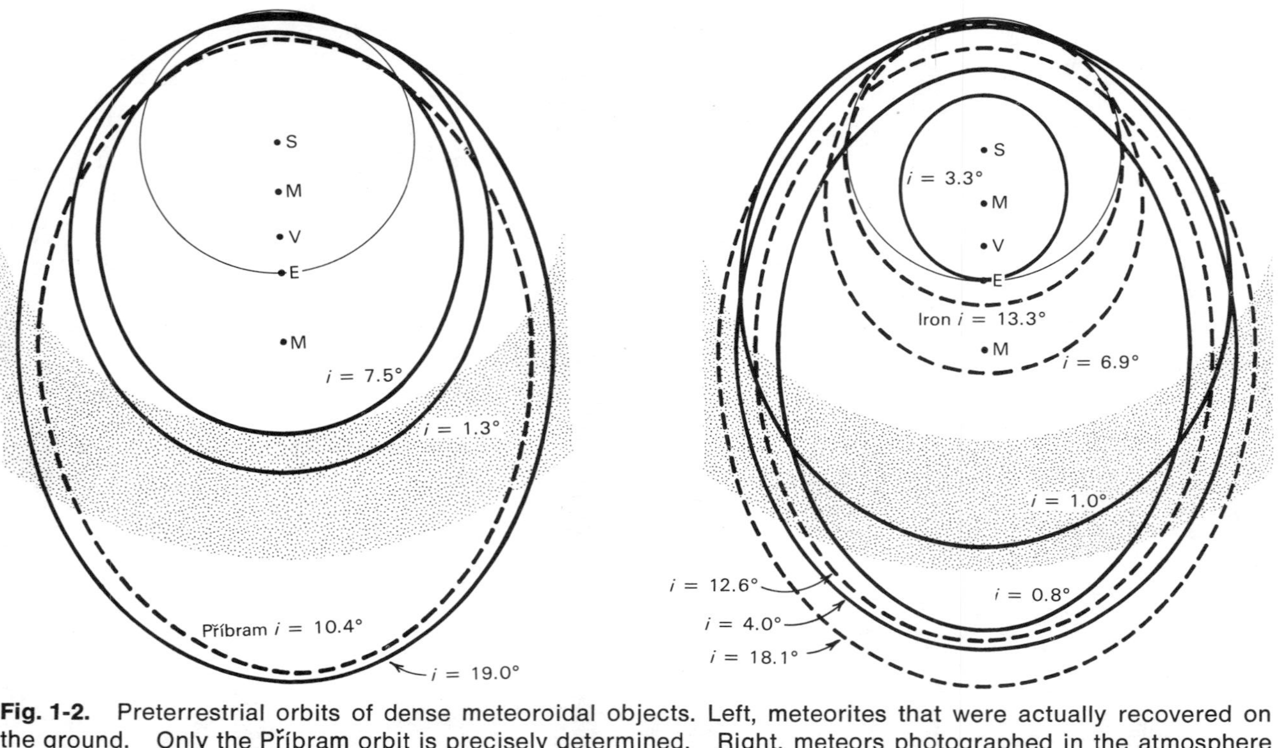

Fig. 1-2. Preterrestrial orbits of dense meteoroidal objects. Left, meteorites that were actually recovered on the ground. Only the Příbram orbit is precisely determined. Right, meteors photographed in the atmosphere but not recovered. Prairie network orbits are indicated by solid lines. All orbits have been drawn with major axes aligned, and in two dimensions. Reader should picture ellipses inclined to the page by the angles *i*. Positions of sun (S), Mercury (M), Venus (V), Earth (E), Mars (M), and asteroid belt (shaded) are indicated.

phere), and these are believed to be identical in composition and source to the larger meteorites. The orbits of four of these that have been discovered appear in Fig. 1-2. The spectrum of one of these meteorites showed it to be composed of iron. It is impossible to recover such tiny objects at the Earth's surface, of course.

Finally, an organized effort has recently been mounted to photograph the fireballs caused by objects large enough to reach the ground intact, so that their former orbits and landing points can be calculated—to do deliberately what was done at Příbram accidentally. This is the Smithsonian Astrophysical Observatory prairie network, a system of 16 automated camera stations dispersed over the Plains States (Fig. 1-3), where the recovery of a fallen meteorite is potentially

Fig. 1-3. Positions of camera stations in the Smithsonian Astrophysical Observatory prairie network.

easiest. Since the system went into operation in 1964, orbits have been obtained for four objects that were probably* meteorites (Fig. 1-2).

* Judging from their size (brightness) and strength (they were solid enough to generate sonic booms in the atmosphere). However, they were not found after they fell.

The orbital information collected in Fig. 1-2 tells us several things. First, all the objects traveled in closed elliptical orbits, demonstrating that meteorites are creatures of the solar system. Visitors from interstellar space would have described parabolic or hyperbolic orbits. Second, the meteorite orbits tend to extend far out in the solar system, usually reaching aphelion in the asteroid belt, between the orbits of Mars and Jupiter. This appears to confirm a theory that was already widely held before these orbits had been determined, namely, that asteroids are the source of meteorites.

METEORITES FROM ASTEROIDS

Asteroids are small, solid bodies, enormous numbers of which orbit between Mars and Jupiter. The largest, Ceres, is 770 km in diameter. There are about 10,000 asteroids larger than 10 km in diameter and perhaps 100,000,000,000,000 (a very crude estimate) larger than 1 meter. In spite of their colossal numbers, however, the mass of all of them put together is only ~3% of that of the moon. In general the asteroids are visible only as faint points of light in the night sky, yet we are able to say that many of them probably have irregular, fragmental shapes. This is so because their brightness varies periodically, typically increasing and decreasing every few hours. It suggests that we are seeing in turn the sunlit faces of spinning, blocky shaped objects—first a broad face, then a narrow edge, then a broad face again, etc.

The orbits of 1,660 asteroids have been accurately determined. The great majority of these remain in the asteroid belt, between Mars and Jupiter. But a few (34 orbits determined) cross the orbit of Mars and penetrate closer in to us, and an even smaller fraction (8 orbits determined) cross the orbits of both Mars and Earth. One of the latter, Hermes, has passed within 800,000 km of the Earth (about twice the Earth-to-moon distance).

It is believed that originally there were far fewer asteroids, perhaps 10 to 100 bodies, and that as time passed these collided with one another and were broken up, and the angular blocks of rubble were scattered about the asteroid belt.

Irregular collision fragments would, of course, appear to vary in brightness as they tumble and spin in space. Some of the larger asteroids we see today may have escaped serious damage, but the smaller boulders and pebbles are almost certainly collision debris, fragments of larger bodies that were demolished ages ago.

The orbits of asteroids (and of planets in general) are not immutable, but are constantly being changed. Collisions deflect them, billiard-ball fashion. Whenever one object passes close by another that is vastly larger, the gravitational field of the latter bends the path or trajectory of the smaller object, changing the character of its orbit. Mars is particularly effective in perturbing the orbits of asteroids in this manner. The Mars- and Earth-crossing asteroids are very probably former belt asteroids which were perturbed into their present orbits by Mars.

The eight known Earth-crossing asteroids must be accompanied by a huge number of smaller fragments, not large enough to be observable. As the orbits of these shift and fluctuate, they must from time to time bring asteroidal objects into collision with the Earth. Such objects, if recovered after the collision, are meteorites, by definition. Thus, at least some of the meteorites that fall must be of asteroidal origin.

METEORITES FROM THE MOON

However, some may have come from a different source. They may be pieces of the moon, broken out and hurled away on occasions when large asteroids or comets struck the lunar surface and blasted craters in it. (It is all but impossible that fragments could have been thrown out from Mars, Venus, or the Earth itself. Rocky material cannot have been accelerated to the escape velocities of these planets without being destroyed by crushing or melting.) H. C. Urey has suggested that most of the meteorites have come from the moon rather than the asteroids. Most of the ejected lunar debris would follow orbits about the sun more or less like that of the Earth-moon system until it was captured, either by Earth or by the moon. It would appear that we could rule out the possibility of a lunar origin for most of the meteorites on this basis, since meteoritic material does not generally arrive in orbits that are circular and Earth-like, but rather in elliptical orbits that extend out into the asteroid belt (Fig. 1-2).

Surprisingly, however, this is not the case. J. R. Arnold, of the University of California (San Diego), has programmed a digital computer to investigate the fate of numerous hypothetical pieces of lunar debris, assumed ejected from the moon by cratering explosions. He finds that most of the fragments would be swept up from their Earth-like orbits very promptly, by Earth or moon. But some of the fragments would be perturbed by the Earth into orbits that carried them far out in the solar system: orbits similar to the Earth-crossing asteroids and the meteorites. Once in these orbits they would be far less susceptible to capture by the Earth and moon than the debris in Earth-like orbits was, and many would orbit for millions of years before capture. Thus one can postulate that major cratering events occur rarely on the lunar surface and that the last of these happened some millions of years ago. It threw a huge amount of debris into space, in Earth-like orbits. Some was perturbed into asteroid-like orbits, which we intercept from time to time, and this is the source of meteorites. The rest was captured long ago, before man's appearance on the scene. No fragments still in telltale Earth-like orbits have survived capture.

So the meteorites, which travel in orbits similar to Earth-crossing asteroids, may have come from the moon, or they may have come from the Earth-crossing asteroids. Although the latter possibility seems somewhat more straightforward, it is not possible at present to exclude either alternative with any degree of certainty.

Whatever their source, meteorites enter the Earth's atmosphere at velocities of 10 to 20 km/sec. Small meteorites are slowed so much by air resistance that they usually do not fracture on Earth impact. But the larger a meteorite is, the less the atmosphere can slow it. A meteorite larger than about 100 tons would be more massive than the volume of air it punches through as it penetrates the atmosphere; thus its velocity would not be checked substantially. If such an object fell, its kinetic energy at impact would be enormous—something like 20,000 joules/g (compare with the chemical energy of nitroglycerine: 6,600 joules/g). It would explode as violently as a nuclear bomb. Fortunately, the Earth very rarely sweeps up such large objects. The only explosive encounter in historical times occurred near the Tunguska River, Siberia, in 1908. Some 35 structures in the Earth have

Fig. 1-4. The Arizona Meteor Crater, a wound in the earth over 1 km in diameter, made by an impacting iron meteorite in prehistoric time. Thousands of meteorite fragments, aggregating over 30 tons, have been found scattered over the surrounding desert. (*U.S. Air Force photograph.*)

been identified with varying degrees of certainty as ancient meteorite explosion craters. Foremost among these is the Arizona Meteor Crater, 1.2 km in diameter and 140 meters deep (Fig. 1-4).

Meteorites are named from the ancient Greek word *mete-*

ōra ("things in the air"). Something like 500 of them (fist-sized or larger) strike the earth each year, according to our best estimates, but only a few—10 or 20 a year—are recovered and ultimately find their way into museum collections. By now these collections contain about 700 meteorites that were actually seen to fall ("falls") and another 900 that were found in the soil ("finds"). Each is given the name of its place of fall or discovery; thus meteoritic jargon includes many wonderfully exotic place-names in obscure languages.

SUGGESTED FURTHER READING ON TOPICS IN CHAPTER 1

1 Atmospheric flight of meteorites, eyewitness reports, impact and cratering, meteorite recovery: H. H. Nininger, "Out of the Sky," 1952 (Dover Publications, Inc., New York, 1959, paperback ed.).

2 The Prairie Meteorite Network: R. E. McCrosky and H. Boeschenstein, S.A.O. Spec. Rept. 173, 1965. Obtainable from Smithsonian Astrophysical Observatory, 60 Garden St., Cambridge, Mass. 02138.

3 Asteroids, their numbers and properties: F. G. Watson, "Between the Planets," chaps. 1–3, 1956 (Doubleday & Company, Inc., Garden City, paperback ed.).

4 Calculated orbits and lifetimes in space of lunar versus asteroidal fragments: J. R. Arnold, The Origin of Meteorites as Small Bodies, *Astrophys. J.,* vol. 141, pp. 1536–1556, 1965.

5 Meteorite craters and the Tunguska River event: B. M. Middlehurst and G. P. Kuiper (eds.), "The Moon, Meteorites, and Comets," The University of Chicago Press, Chicago, 1963, contains several informative chapters (7–11) on these topics.

6 The frequency of meteorite impacts on the Earth: G. S. Hawkins, Asteroidal Fragments, *Astron. J.,* vol. 65, pp. 318–322, 1960.

2
WHAT ARE THEY MADE OF?

> The outside of every stone that has been found, and has been ascertained to have fallen from the cloud near Sienna, is evidently freshly vitrified, and is black, having every sign of having passed through an extreme heat; when broken, the inside is of a light-gray color mixed with black spots, and some shining particles, which the learned here have decided to be pyrites. . . .
>
> W. Hamilton (1795)*

As people began to take more seriously the proposition that certain stones had fallen from the heavens, they began to examine these stones more closely. The first (1802) careful description of the physical nature of meteoric stones was made by Edward Howard, an English chemist, and Jacques Louis de Bournon, a French nobleman and former army officer exiled in England after the revolution. They obtained samples of the stones that had recently fallen at Sienna (Italy) and Wold Cottage (England), and also meteorite specimens from India and Bohemia.

* From Account of a Fall near Siena, *Phil. Trans. Roy. Soc., London,* vol. 85, pp. 103–105.

All four samples were found to be quite similar. They consisted of gray, granular, earthy material. Sprinkled about in this, several different kinds of particles could be discerned. One was the shining yellowish crystals noted by Hamilton, above, and thought to be pyrite (FeS_2). But De Bournon observed two other kinds:

> One of these substances, which is in great abundance, appears in the form of small bodies, some of which are perfectly globular, others rather elongated or elliptical. They are of various sizes, from that of a small pin's head to that of a pea, or nearly so: some of them, however, but very few, are of a larger size. The color of these small globules is gray, sometimes inclining very much to brown: and they are completely opaque. They may, with great ease, be broken in all directions: their fracture is conchoid, and shows a fine, smooth, compact grain, having a small degree of lustre, resembling in some measure that of enamel. Their hardness is such, that, being rubbed upon glass, they act upon it in a slight degree; this action is sufficient to take off its polish, but not to cut it: they give faint sparks, when struck with steel.
>
> [The other] consists in small particles of iron, in a perfectly metallic state, so that they may easily be flattened or extended by means of a hammer. These particles give to the whole mass of the stone, the property of being attractable by the magnet; they are, however, in less proportion than those of pyrites just mentioned. When a piece of the stone was powdered, and the particles of iron separated from it, as accurately as possible, by means of a magnet, they appeared to compose about $2/100$ of the whole weight of the stone.*

Howard isolated a number of these metal grains and analyzed them chemically. He found that they were a mixture or alloy of iron with nickel. Nickel-iron and the small globular bodies noted by De Bournon occur only in meteorites (Fig. 2-1) and serve to set them apart from all known terrestrial rocks.

* From E. Howard, Experiments and Observations on Certain Stony and Metalline Substances, Which at Different Times Are Said to Have Fallen on the Earth; Also on Various Kinds of Native Iron, *Phil. Trans. Roy. Soc. London,* vol. 92, pp. 168–212, 1802.

Fig. 2-1. A piece of the Ankober (Ethiopia) chondrite. A sawed surface has been smoothed and polished, so that internal structure is visible. Evenly dispersed, irregular white flecks are highly reflective grains of Ni–Fe metal and Fe sulfide. Inconspicuous rounded structures in varying shades of gray are chondrules. All are embedded in a gray silicate matrix. (*Smithsonian Astrophysical Observatory photograph.*)

TYPES OF METEORITES

There are, of course, various kinds of meteorites: stones, irons, and stony-irons.† But as more and more meteorites were collected by the great museums of the world, it became apparent that the great majority of them—about 85% of those seen to fall—were stones similar to those described by Howard and De Bournon. This may come as a surprise to some readers. Most people think of meteorites as huge masses of nickel-iron metal, because this is what they see in museum exhibits. Museum collections contain inordinately many iron meteorites because irons are much more likely to be found than stony meteorites. Irons are conspicuously heavy, while to the untrained eye stony meteorites look just like terrestrial rocks. Also, museum exhibits tend to favor irons because

† In the older literature these are termed *aërolites, siderites,* and *siderolites.*

they are bigger and more awe-inspiring than stones. But if we base our statistics only on those meteorites that were actually seen to fall, and exclude those accidentally found on or in the soil, it turns out that irons are rather rare (Fig. 2-2).

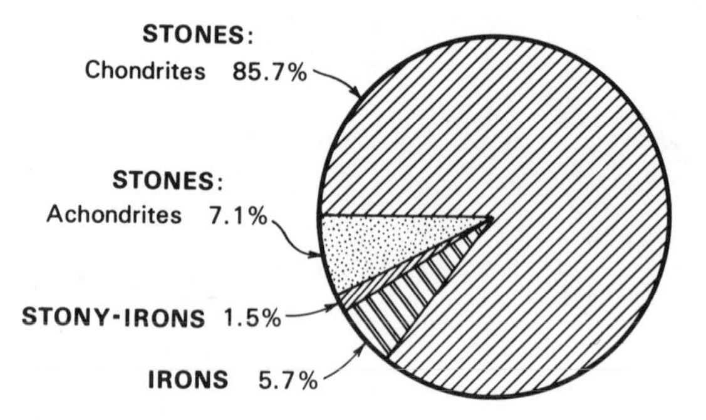

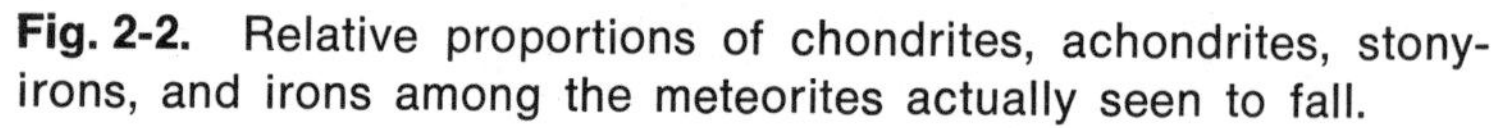

Fig. 2-2. Relative proportions of chondrites, achondrites, stony-irons, and irons among the meteorites actually seen to fall.

When Gustav Rose cataloged the meteorite collection of the University of Berlin in 1864, he named the type of meteorite Howard and De Bournon had studied the *chondrites.* He based this name on the ancient Greek word *chondros,* meaning "grain of seed," a reference to the tiny rounded bodies that characterize these stones. The little spheroids themselves have come to be called *chondrules.*

Not all stony meteorites are chondrites. A relatively small number of them are quite different from the chondrites in composition and structure. They are for the most part quite similar to terrestrial igneous rocks and contain no chondrules or nickel-iron metal. These stones are referred to collectively as *achondrites.*

CHONDRITES SEEN THROUGH THE MICROSCOPE

Let us focus attention on the principal class of meteorites, the chondrites. We come now to the contribution of a very remarkable man, Henry Clifton Sorby. Two names belong at the head of the list of people who have concerned themselves with meteorites: Chladni and Sorby. Sorby, a bachelor Englishman, devoted his long life to researches in various scientific fields, chiefly geology. He was independently wealthy, in the best tradition of nineteenth-century gentleman scholars.

Optics was another of Sorby's interests, and it occurred to him early in his life that geology would profit if rocks could be studied under the microscope. He realized that immeasurably more would be learned if, instead of simply pointing the microscope at the rough broken surface of a rock, one could take a very thin, smooth slice out of the rock and examine it microscopically while light was shined up through it. In this way one could literally see *into* the rock and study in infinite detail the tiny mineral crystals and the way they fitted together to make up the rock. In 1849, when Sorby was only twenty-three, he prepared the first such rock thin section, opening up a whole new area of scientific study: microscopic petrography. Sorby's thin sections were made by a laborious process of hand grinding. A flat surface was ground on each specimen, and against this a glass slide was cemented with clear Canada balsam. The remaining rock was then ground away until only a layer of about 0.03 mm thickness remained, cemented against the glass. Silicate minerals ground this thin are quite transparent. (Thin sections are still made today by a mechanized form of the same process.)

In the 1860s, after studying countless thin sections of terrestrial rocks, Sorby turned his attention to the chondrites. He was the first to see the complex internal structure of these stones (Fig. 2-3) and the striking and varied patterns of crystals within the chondrules. He recognized immediately that the chondrules are igneous in character—that they were once hot droplets of melted rock.

> . . . the form and structure of many of the grains is totally unlike that of any I have ever seen in terrestrial rocks, and points to very special physical conditions. Thus some are almost spherical drops of *true glass* in the midst of which crystals have been formed, sometimes scattered promiscuously, and sometimes deposited on the external surface, radiating inwardly; they are, in fact partially devitrified globules of glass, exactly similar to some artificial blow-pipe beads.
>
> The nearest approach to the globules in meteorites is met with in some artificial products. By directing a strong blast of

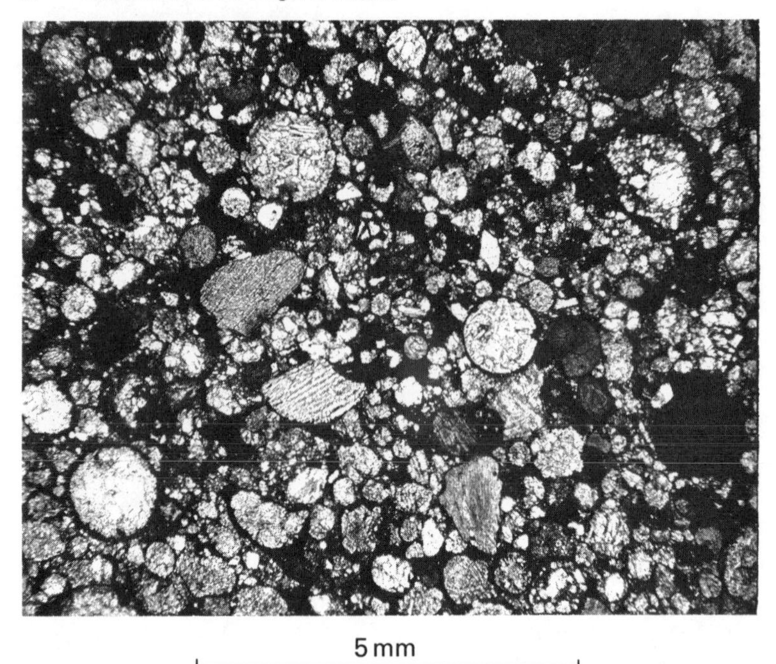

Fig. 2-3. Microscopic view of a thin section of the Tieschitz (Czechoslovakia) chondrite. Relatively transparent minerals appear light; opaque substances are darker. It can be seen that the stone is an aggregation of rounded structures; these are chondrules, once droplets of a "fiery rain."

hot air or steam into melted glassy furnace slag, it is blown into spray. . . . If the slag be hot enough, some spheres are formed without tails, analogous to those characteristic of meteorites. . . . . . . melted globules with well-defined outlines could not be formed in a mass of rock pressing against them on all sides, and I therefore argue that some at least of the constituent particles of meteorites were originally detached glassy globules, like drops of a fiery rain.*

Conditions on the surface of the sun seemed to Sorby almost exactly what would be needed to produce a fiery rain

* From On the Structure and Origin of Meteorites, *Nature,* vol. 15, pp. 495–498, 1877.

of chondrules, and he suggested that chondrites might well be pieces of the body of the sun, cast out by solar prominences. Alternatively, they might be agglomerations of

> . . . residual cosmic matter, not collected into planets, formed when conditions now met with only near the surface of the sun extended much further out from the center of the solar system.†

One cannot help admiring the sharpness of Sorby's eye and his ability to grasp the meaning of what he saw, especially now that, a century later, we are rediscovering the things he talked about. He has to be forgiven the naïve idea about chondrites being pieces of the sun: during the first half of the nineteenth century Sir John Herschel's conception of the sun as a solid, cool Earth-like body covered over with a relatively thin layer of white-hot gases was generally accepted. Not until the 1860s, at the time Sorby was making his observations, did astronomers begin to think of the sun as we do today, as a wholly gaseous, fluid body.

In passing it should be noted that Sorby, in order to extend his microscope studies to the iron meteorites, founded yet another branch of science, metallography.

THE CHEMICAL COMPOSITION OF CHONDRITES

Nothing has been said yet about the composition of chondrites. Rock analysis is a difficult business, and that of chondrites is doubly so because they consist not only of rock, but also of sulfide and metallic minerals. The first crude partial analyses of chondrites were made by eighteenth-century chemists, including the famous Antoine Lavoisier, but not until about 1870 did reasonably accurate analyses begin to appear. Even today the analysis of chondrites by wet chemical methods is considered more of an art than a science, and currently the work of only a handful of people, foremost among them H. B. Wiik, of the Geological Survey of Finland, is so artistic as to be acceptable without question to

† *Ibid.*, pp. 495–498.

everyone. Wiik's analysis of the Richardton, North Dakota, meteorite, a typical chondrite, follows:

Stony material	wt %
SiO_2	34.3 wt %
MgO	22.2
FeO	9.9
Al_2O_3	2.6
CaO	1.4
Na_2O	1.0
Cr_2O_3	0.6
P_2O_5	0.5
H_2O	0.5
MnO	0.4
C	0.2
K_2O	0.1
TiO_2	0.1
Total stony material	73.5

		wt %
	Fe	18.3 wt %
	Ni	1.6
	Co	0.1
Metallic grains		20.0
	FeS	6.0
Sulfide grains		6.0

The analysis of the gray stony material that makes up most of the chondrite is presented here in terms of oxides rather than chemical elements (that is, SiO_2 instead of Si). This is customary in rock and ceramic analyses. In rocks the various metallic elements do occur fully combined with oxygen, but usually in complex compounds (minerals) involving several different metallic elements, not as simple oxides.

Presenting the analysis in this way points up the fact that the element iron occurs in three different forms in a chondrite. That part reported as "FeO" resides in silicate minerals; the "Fe" resides in grains of nickel-iron; and some iron occurs in sulfide minerals (FeS). However, there are other ways of couching the analysis. For example, we could eliminate all O, S, C, and H from consideration and simply list the relative amounts of the various metallic elements present. In this case all iron, whether silicate, metal, or sulfide, would be included in a single entry.

As more and more good analyses of chondrites became available in the nineteenth century, it became evident that their compositions (on a metals-only basis, as just discussed) are very similar to one another, perhaps identical. They seemed to differ only in the degree of oxidation of their iron. Some contained much metallic Fe but little FeO (hence iron-poor silicate minerals). In others the situation was reversed: little Fe metal, but much FeO. This relationship was first noted by the Swedish chemist A. E. Nordenskjöld in 1878, on the basis of superior analyses of nine chondrites. Nordenskjöld concluded that all chondrites must have been derived from a common parent material of uniform composition, in which all iron was at first fully oxidized. In different portions of this material the FeO came to be reduced* to varying degrees, so that in some cases most of it was turned to metal while in other cases most of it remained oxidized. This has proved to be a very attractive idea; a number of writers have adopted it in the century since Nordenskjöld.

Actually we know now that all chondrites are not identical in composition and could not be derived from a common parent material by simple reduction. Not only the degree of oxidation, but also the absolute abundance of iron is variable among the chondrites. Improved analyses have made this apparent. Figure 2-4 illustrates the differences. The ratio (Fe present as metal)/(total Fe present) reflects the degree of oxidation of the iron, and here we see the variability pointed out by Nordenskjöld. The ratio Fe/Si is a measure of the amount of iron present (on a metals-only basis), and here we see more fundamental differences that cannot be changed by reduction or oxidation.

Note that points representing individual chondrites are not dispersed at random in Fig. 2-4, but instead fall into several well-defined clusters. These groups will be further discussed in Chap. 3.

Some of the groups have the same average values of Fe/Si as others, and it might seem that the C and H groups could have been derived by reduction from one parent material

* *Reduction* means (for our purposes) the taking away of oxygen from a compound. The commercial smelting of iron is a prime example of reduction:

Fe_2O_3	+	3C	→	2Fe	+	3CO
(iron ore)		(coal)		(pig iron)		(carbon monoxide, driven off)

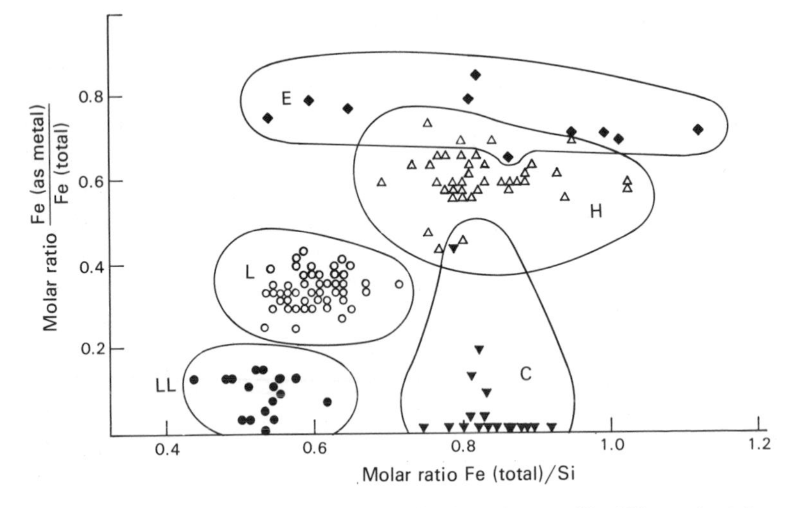

Fig. 2-4. Differences in the overall abundance (Fe/Si) and state of oxidation (Fe metal/total Fe) of Fe in chondrites: data from bulk chemical analyses of 130 chondrites. On the basis of these and other chemical differences, chondrites can be separated into five groups (circled).

while the LL and L groups came from another. Even this appears impossible, however, because of other small differences in metals-only composition (such as Mg/Si, Na/Si) between the groups. It seems that each chondrite group is representative of a fundamentally different batch of planetary material. All these very important aspects of chondrite chemistry were discovered by H. C. Urey and H. Craig, then at the University of Chicago.

THE UNDIFFERENTIATED CHARACTER OF CHONDRITIC METEORITES

So far we have talked only about the 15 or so most abundant elements in chondrites. The first comprehensive study of the rarer elements was carried out in 1930 by the German chemists Walter and Ida Noddack (husband and wife). The Noddacks measured the abundances of a great many trace elements by methods of optical (arc) and x-ray spectrography. Trace-element analysis is another difficult business, and some of the values they obtained have not stood the test of time as well as others have. Yet the general conclusion they arrived at was quite correct, and extremely important. This was that the meteorites have a more *generalized* composition than the crust of the Earth; they contain more different elements (lithophile, chalcophile, and siderophile), at levels detectable by the Noddack's 1930 techniques, than Earth rocks.

Lithophile elements (such as Se, Sr, Rb, Ba, Ce, Cs, Th, U) are those with a great affinity for minerals rich in oxygen; chalcophile elements (such as Cu, Zn, Sn, Pb, Ag, Hg, Cd, In) have a similar affinity for sulfur-bearing minerals; and siderophile elements (such as Ge, Ga, Ru, Pt, Pd, Os, Ir, Rh) prefer to be wherever metallic iron and nickel are. Natural processes that operate in and on the earth—weathering, sedimentation, metamorphism, melting, etc.—have a very strong tendency to separate or differentiate these three element groups from one another, so that we find the rocks of the Earth's crust generally composed of lithophile elements, while the chalcophile elements are highly concentrated in a few small areas (ore deposits) and the siderophile elements are largely missing from the crust. Presumably the latter are concentrated in the core of the Earth, which is thought to consist of nickel-iron metal. But in chondritic meteorites, lithophile, chalcophile, and siderophile elements occur mixed together to a degree not matched in terrestrial rocks: it would appear that chondrites have not been chemically differentiated the way the Earth's rocks have, so they cannot have suffered through the processes that cause differentiation (melting, etc.), and therefore they must have survived essentially unchanged since the time when the planets formed.

The undifferentiated character of chondritic meteorites can best be illustrated by comparing their composition (chemically determined) with the composition of the sun (from spectographic studies of the solar atmosphere). The sun and planets are thought to have formed from a single fairly homogeneous cloud of gas and dust (Chaps. 6 and 7), so planetary matter, if it has remained undifferentiated, should still show the same relative abundances of *condensable* elements* as the sun. The comparison was first made in

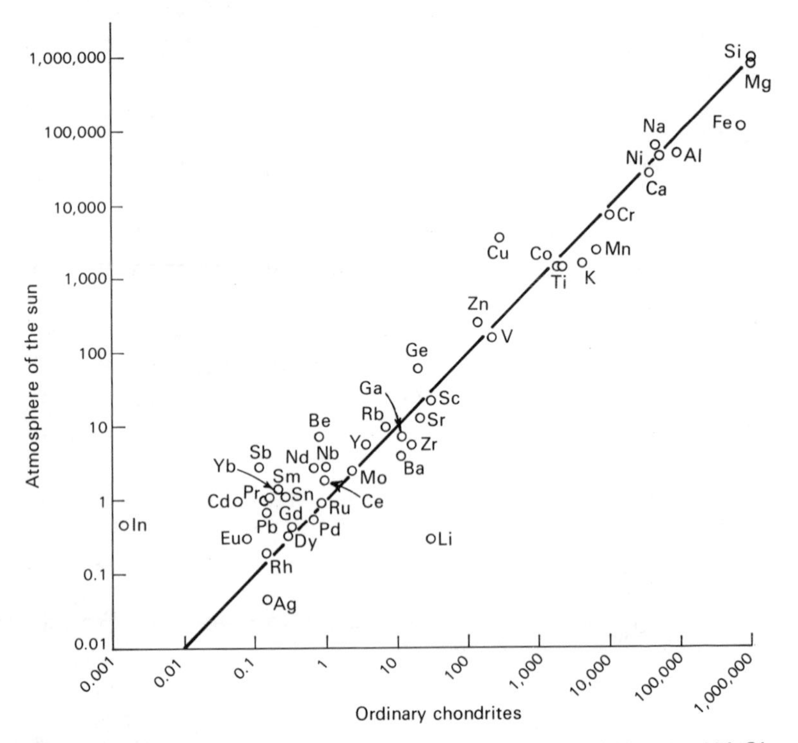

Fig. 2-5. Abundances of the metallic elements (relative to 10^6 Si atoms) in ordinary chondrites, compared with abundances in the sun. The correspondence is good but not perfect. Many of the abundance discrepancies that do exist can be rationalized (Chaps. 6, 7). (Cu in the sun is poorly determined.) Principal paradox is Fe.

* We have to exclude elements with very low boiling points—H, He, O, N, etc.—from the comparison. Temperatures in the inner solar system are not likely to have been low enough to condense these at the time the planets came together. Even if they had been, present-day temperatures near and on the Earth would not have allowed chondrites to retain them in abundances comparable to the sun.

1929 by Henry Norris Russell, of the Mount Wilson Observatory, when he obtained from spectograms the first quantitative analysis of the solar atmosphere. Russell found the sun tolerably close to chondrites in composition—much closer to chondrites than to terrestrial rock compositions.

The comparison as it stands today is shown in Figs. 2-5 and 2-6. Relative solar abundances in these are mostly from the recent measurements of Goldberg, Aller, and Müller, then of the University of Michigan. Entries are made for the 44

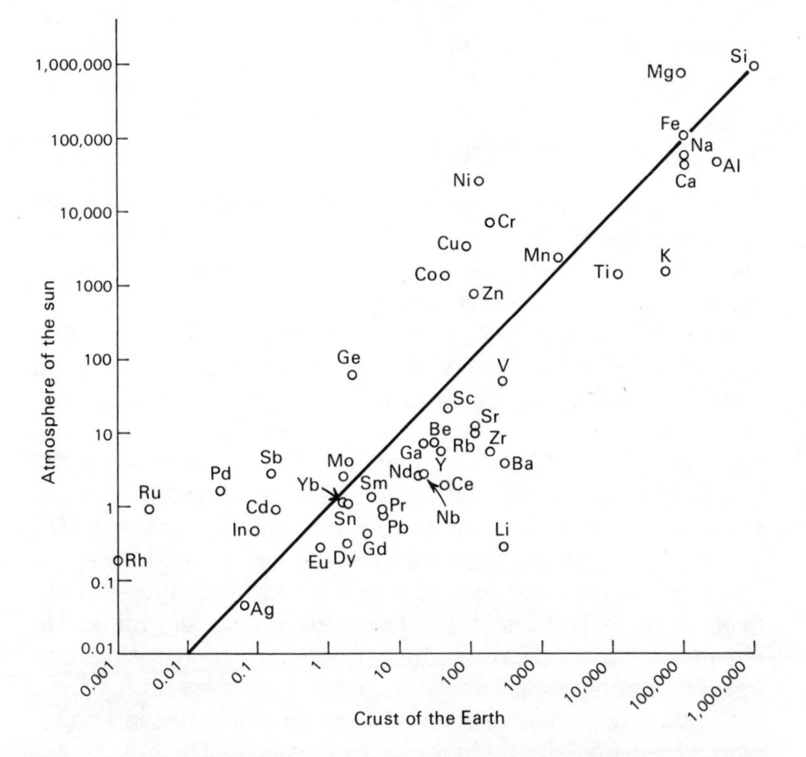

Fig. 2-6. Metal abundances in the Earth's crust (a differentiated rock system) versus the sun. Here we see a tendency for siderophilic and chalcophilic elements to be depleted in the crust; they plot above the 45° line. Lithophilic elements are enriched in the crust and plot below the 45° line. (Crustal abundance estimates by S. R. Taylor.)

condensable elements whose solar abundances have been calculated. In Fig. 2-5 the sun is compared with chondrites; in Fig. 2-6 with the mean composition of the Earth's crust. The elemental abundances shown for chondrites and crustal rocks are based on the most up-to-date measurements, many made by methods of neutron-activation analysis. All abundances are calculated relative to an assumed 1,000,000 atoms of Si. We see that, indeed, chondritic abundances match the solar values much more closely than do abundances in the Earth's crust. The correspondence between chondrites and the sun is, with a few exceptions, very good. The exceptions will be touched upon later.

MINERALS IN CHONDRITES

Finally, what about the compounds or minerals that make up chondrites? Chemical analysis tells us about elemental abundances but gives no explicit information as to the manner in which the various elements are combined with one another. The high SiO_2 content of chondrites insures that they, like terrestrial rocks, must consist largely of silicate minerals. Mineralogical studies have shown that they consist dominantly of olivine and orthopyroxene. (Olivine is a solid-solution mixture of fayalite, Fe_2SiO_4, with Mg_2SiO_4. Mixtures in any proportion are possible. Similarly, orthopyroxene is a mixture of ferrosilite, $FeSiO_3$, with $MgSiO_3$. Olivine and orthopyroxene compositions are expressed in terms of the mole percent of fayalite and ferrosilite, respectively, present.) Most of the transparent mineral grains visible in a chondrite thin section (Fig. 2-3), both inside and between chondrules, consist of these two minerals.

Chondrites contain several other silicate minerals in smaller amounts, especially feldspar and diopside. They also contain metallic nickel-iron minerals (taenite and kamacite), as noted by Howard, and a sulfide mineral. The latter is troilite (FeS), however, not pyrite (FeS_2), as Hamilton and De Bournon thought. The proportions of various minerals present

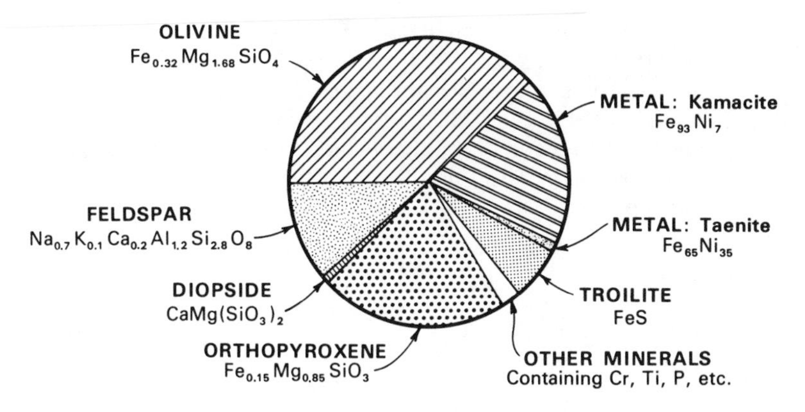

Fig. 2-7. Mineral content of a typical chondrite, Richardton. Areas in diagram are proportional to weight percentages.

in a typical chondrite (Richardton) are summarized in Fig. 2-7. All these except the metallic minerals also occur in abundance in the Earth's crust.

SUGGESTED FURTHER READING ON TOPICS IN CHAPTER 2

1 Detailed descriptions of structures and minerals in chondrites: the best thin-section photographs are still those assembled by G. Tschermak, in The Microscopic Properties of Meteorites (1883); this is now available as *Smithsonian Contrib. Astrophys.,* vol. 4, no. 6, 1964. See also P. Ramdohr, The Opaque Minerals in Stony Meteorites, *J. Geophys. Research,* vol. 68, pp. 2011–2036, 1963.

2 Chemical composition of ordinary chondrites: content and variability of major elements is detailed in B. Mason, The Chemical Composition of Olivine-bronzite and Olivine-hypersthene Chondrites, *Museum Novitates* 2223, 1965. Trace-element abundances and their implications are discussed in J. W. Larimer and E. Anders, Chemical Fractionations in Meteorites. II. Abundance Patterns and Their Interpretation, *Geochim. Cosmochim. Acta,* vol. 31, pp. 1239–1270, 1967.

3 The concept that all chondrites are derived from a single highly oxidized parent material: Nordenskjöld's idea is further developed in A. E. Ringwood, Chemical and Genetic Relationships among Meteorites, *Geochim. Cosmochim. Acta,* vol. 24, pp. 159–197, 1961.

4 Solar abundances of the elements: L. Goldberg, E. A. Müller, and L. H. Aller, The Abundance of the Elements in the Solar Atmosphere, *Astrophys. J. Suppl. Ser.,* vol. 5, no. 45, 1960.

3
THE PARENT METEORITE PLANETS

[I have concluded that] *the meteorites are pieces of debris from a disrupted planet.* Now just as one can, from exhumed remains of extinct animals, reconstruct the beings of past epochs, so it should be possible by examining meteorites to reconstruct the celestial body that supplied these fossil vestiges. . . .

At the center of the globe there was, evidently, a core of meteoric iron. . . . Above it was probably iron containing grains of olivine, as in the well-known mass discovered at Krasnojarsk and described by Pallas. Next would come the true *stones,* at first those containing large nodules of metal, like the meteorite of Sierra de Chaco; then ones with finer and finer metal grains, such as the masses from L'Aigle, Aumale, Lucé and Montrejeau; and finally stones absolutely devoid of metal, for example the Chassigny and Juvinas falls.

M. S. Meunier (1871)*

Have meteorites always been small lumps and crumbs of earthy matter wandering through space—leftovers from the

* From Structure du globe d'oú proviennent les météorites, *Compt. rend.,* vol. 72, pp. 111–114.

act of creation, as it were—or are they fragments broken out of much larger bodies that once existed? Since the time of Chladni the latter view has been almost universally held. The evidence in favor of "parent meteorite planets" is very strong. Especially persuasive are the iron meteorites, so let us look at these in some detail.

IRON METEORITES

The irons consist almost entirely of massive nickel-iron metal. They are rich in siderophile elements and almost devoid of lithophile elements. How can such objects have formed? Their composition is highly specialized, totally dissimilar to the abundance pattern of condensable elements in the sun (Fig. 2-5), so they cannot be samples of primordial solid matter as it first condensed in the solar system. They must instead be the products of some process that chemically fractionated material after it had condensed.

Only melting seems capable of having done the job. Suppose the first solid material that existed in the solar system had a generalized composition, and that some of its nickel and iron were present as metallic alloys and sulfide minerals, as we have seen is the case in the chondritic meteorites (Chap. 2). If such a material were melted, it would not have yielded a single homogeneous fluid. Instead the silicate and oxide compounds would have mingled to form one liquid or magma, while the metals and sulfides would have formed another, immiscible with the first. The two liquids would have remained separated, like oil and water. If melting occurred within a sizable accumulation of primitive planetary matter, massive enough to generate a perceptible gravitational field, then the denser (heavier) metal-sulfide liquid would have sunk to the center of the object and coalesced into a molten core. The lighter silicate magma would have surrounded and floated on top of the core. (The commercial smelting of iron and other metals depends on this tendency of metallic and silicate melts to separate.) The liquid at the core would have been similar in composition to iron meteorites. Cooling and crystallization of the system, followed by a shattering collision in space, could have left orbiting core fragments that had the properties of iron meteorites.

Most iron meteorites are internally structured in a curious and striking way. When they are sliced, polished, and etched in dilute acid, the structure becomes visible (Fig. 3-1). It has

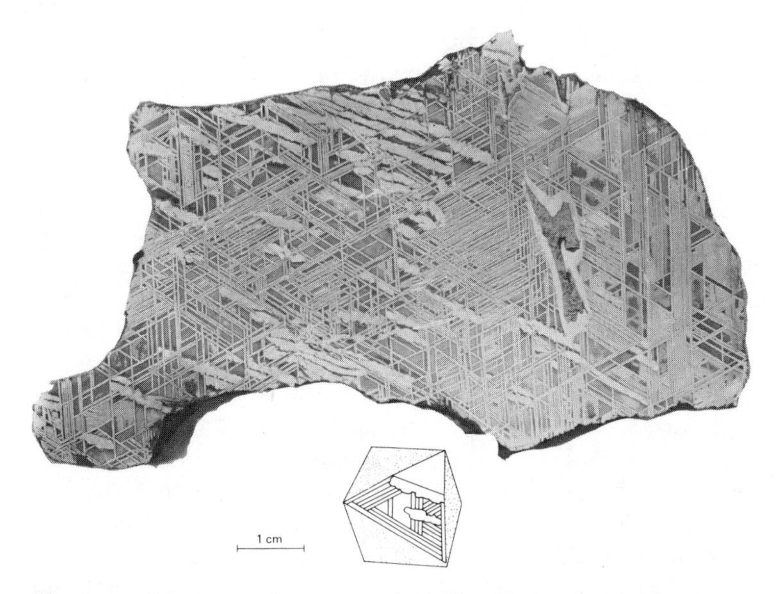

Fig. 3-1. Windmanstätten pattern in the Edmonton (Kentucky) iron meteorite. Systems of bands are kamacite lamellae, which parallel faces of a reference octahedron (sketched). Specimen surface cuts the reference octahedron almost parallel to one face. Irregular inclusion at right center of meteorite slab consists of troilite (FeS). *(Smithsonian Astrophysical Observatory photograph.)*

been named after Count Alois de Widmanstätten, director of the Imperial Porcelain Works in Vienna, who first observed it in an iron meteorite in 1808. The Widmanstätten structure consists of parallel arrays of plates of the low-Ni (6 to 7%) alloy kamacite. Each meteorite contains four such arrays, intersecting with one another in a complex way. The systems run parallel to the four planes defined by the faces of an imaginary octahedron (Fig. 3-1), and for this reason iron meteorites that display the Widmanstätten pattern are called *octahedrites.*

The spaces between kamacite plates are filled with higher-Ni substances: the alloy taenite (30 to 50% Ni), and plessite, which is usually a very-fine-grained mixture of kamacite and taenite. The distribution of Ni across the Widmanstätten pattern is best demonstrated by making electron-microprobe* traverses (Fig. 3-2). These show that Ni content in the zones between kamacite plates is always inhomogeneous in a characteristic way: Ni concentration is highest immediately adjacent to the kamacite plates and sags to lower values between them, forming roughly M-shaped profiles.

Although octahedrites cannot be made in the laboratory (the processes involved take too long), we can understand theoretically how it was done—how the Widmanstätten structure and M-shaped Ni distributions must have been generated in them. Metallurgical experience tells us what would have happened in a mass of molten nickel-iron metal, if it were cooled slowly and steadily. The metal would have crystallized after it cooled beneath (roughly) 1400°C. During hundreds of degrees of cooling thereafter it would have existed in the form of a single homogeneous alloy: taenite. But beneath about 900°C, the situation is not so simple. The phase diagram of the Fe–Ni system (Fig. 3-3) shows which alloy or alloys should be present, if equilibrium prevails, at various temperatures and assuming various concentrations of Ni in the metal mass. We see at the highest temperatures a large field in which homogeneous taenite is the stable alloy, as already noted. And there is a field to the left, at very low

* The electron microprobe is a recently developed and very powerful analytical tool. A polished specimen is placed in a vacuum chamber and high-energy electrons are focused into a tiny spot (~0.001 mm) on it. The electrons cause every element present at the target spot to emit x-rays of a characteristic wavelength. The various wavelengths of x-rays produced are separated by a bent-crystal spectrometer; then the x-ray intensity at each wavelength can be measured and the concentration of the element emitting it can be deduced. Thus the chemical composition of minute volumes of material can be determined. The sample can be viewed with an optical microscope and moved about inside the vacuum chamber, so that any area of interest can be placed under the electron beam.

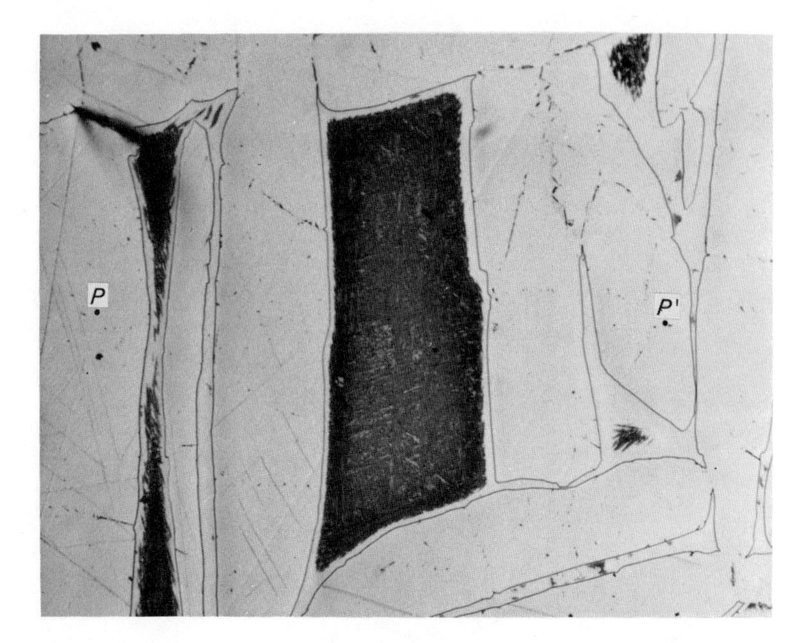

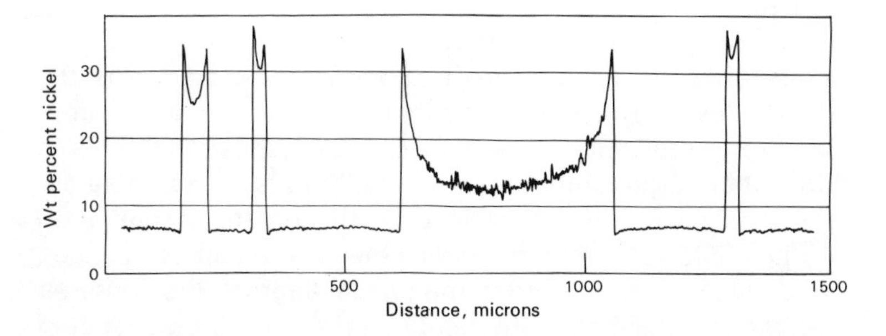

Fig. 3-2. (Above) Microscopic view of Widmanstätten structure in the Anoka octahedrite, viewed by reflected light. Dark plessite areas thinly rimmed by taenite; separated by broad, light-colored lamellae of kamacite. (Below) Electron microprobe profile across the structure (from *P* to *P'*), showing variation in Ni content. (*Figures from J. A. Wood, The Cooling Rates and Parent Planets of Several Iron Meteorites,* Icarus, *vol. 3, pp. 429–459, 1964; courtesy of Academic Press Inc., New York.*)

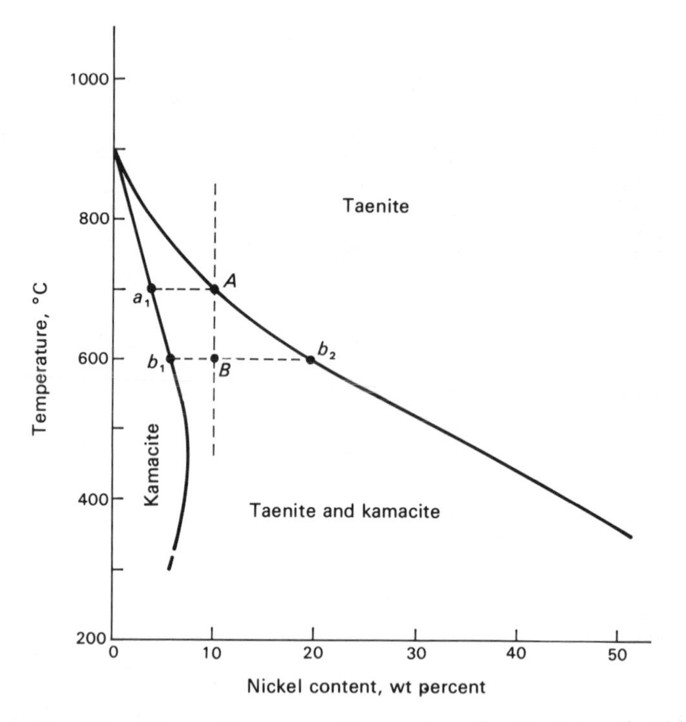

Fig. 3-3. Phase diagram of the system Fe–Ni, beneath 1000°C, at 1 atm pressure.

values of Ni content, where homogeneous kamacite is the stable alloy. (The crystal structures of taenite and kamacite are somewhat different.) Between these fields is a third, where equilibrium requires that both taenite and kamacite should be present. The bulk Ni content of the octahedrites (6 to 15%) is such that as they cooled and so passed vertically downward through the phase diagram, they entered this taenite-plus-kamacite field. Here the Widmanstätten structure must have developed.

Consider a mass with 10% Ni content, cooling along the line *AB* in Fig. 3-3. It passes from the taenite field into the taenite-plus-kamacite field at 700°C. At this point kamacite crystals must begin to appear in the mass, if equilibrium is maintained. Apparently when they did appear in the octahedrites, it was in the form of thin sheets or plates that preferred to grow through the original taenite crystal in a few very special directions, namely, parallel to the {111} lattice planes of the taenite. These planes bear an octahedral relationship to one another, so we can see how the peculiar geometry of the octahedrites was established.

The phase diagram tells us what the Ni content of these first-formed kamacite plates will be. If kamacite and taenite are both present and at equilibrium, their compositions will lie on the boundaries of the kamacite-plus-taenite field. Thus at 700°C equilibrium kamacite has composition a_1 in Fig. 3-3, about 4% Ni, while the taenite still contains ~10% Ni (*A*).

With further cooling, we can see from the sloping field boundaries that these alloy compositions must change. By the time 600°C is reached, equilibrium kamacite must contain ~5.5% Ni (b_1), taenite ~17% Ni (b_2). The Ni content of taenite tends to increase indefinitely with lowering temperature, while kamacite increases its Ni down to ~500°C but tends to lose Ni below this temperature.

How can both of the alloys present increase their Ni contents, if the metal mass as a whole must keep its Ni content constant (at 10%, in the example assumed)? This is possible only if the amount of low-Ni alloy present increases at the expense of the high-Ni alloy. Thus with cooling, the kamacite plates must expand in thickness and the zones of taenite between them must shrink by an equivalent amount.

But how can crystals of solid metal increase their Ni content? Where does the additional Ni actually come from, and how does it get into the kamacite and taenite crystals? In effect, it comes from the interfaces where kamacite and taenite abut against one another, as shown in Fig. 3-4. From there it moves into the alloy crystals by *lattice diffusion.* (The atoms of a crystal perpetually vibrate, and from time to time adjacent atoms are so strongly vibrated that they exchange positions. A Ni atom, introduced at the surface of a pure Fe crystal, can eventually make its way into the crystal by undergoing successive exchanges with Fe atoms. The process is temperature-dependent: the hotter the crystal, the more violently its atoms vibrate, the more often they exchange, and so the greater is the *coefficient of diffusion.*)

At high temperature, diffusion of Ni in taenite occurs so

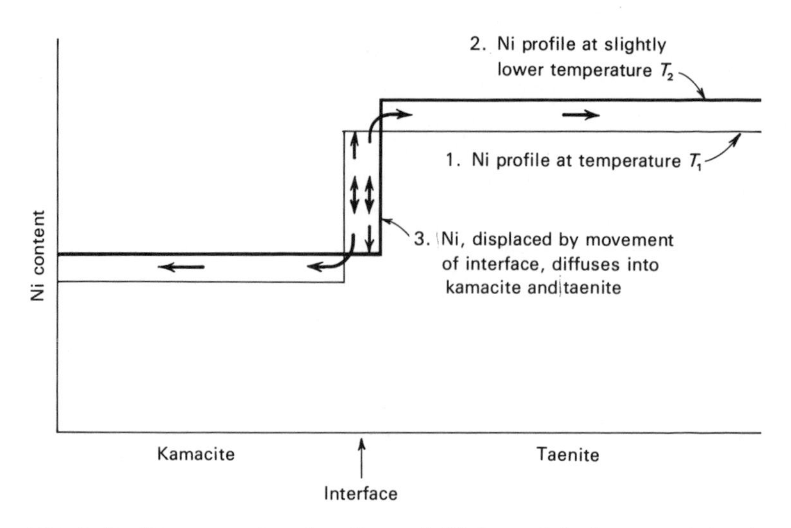

Fig. 3-4. Diagram showing flow of Ni from interface regions to interiors of metal crystals, in cooling octahedrites.

easily that taenite crystals in cooling octahedrites may have had no difficulty in maintaining the equilibrium Ni concentration throughout. As quickly as lowering temperature caused an increase in Ni content at the interfaces, diffusion carried Ni to the centers of the taenite crystals and brought them up to the same concentration. But as the temperature fell lower and lower, and the coefficient of diffusion decreased, a point must have been reached where this was no longer the case. Ni could still diffuse part of the way into a given taenite crystal, but not all the way to its center. The Ni content of the center could not increase after this time. As the temperature fell further, the effective range of diffusion grew shorter and shorter. The final result must have been taenite crystals with higher Ni concentrations near the interfaces than in the interiors—M-shaped Ni distributions like those shown in Fig. 3-2. It appears that the very low-Ni taenite interiors, which were badly out of equilibrium at low temperatures, tended to decompose into a fine-grained mixture of equilibrium taenite and kamacite (plessite).

Ni diffusion is known to occur more easily in kamacite than

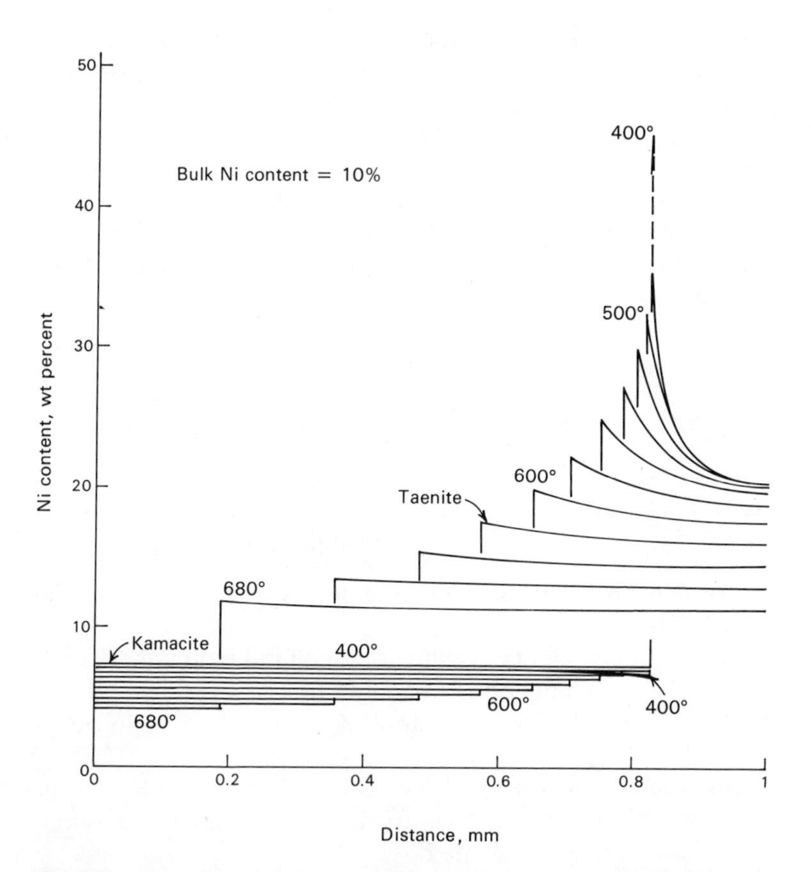

Fig. 3-5. Development of Ni profiles in a cooling octahedrite (computed). Left edge of figure represents midpoint of a kamacite plate; kamacite expands and increases in Ni content as temperature falls. Right edge of figure is center of a residual taenite area. Ni content also increases in shrinking, cooling taenite—at first throughout the taenite, later only at edges. Final result is an M-shaped profile (only half shows in the figure) like those that characterize octahedrites.

in taenite. Evidently diffusion in the meteoritic kamacite crystals did not begin to fail until the temperature had fallen beneath ~500°C, into the portion of the phase diagram where the boundary between the kamacite field and kamacite-plus-taenite field reverses its slope. This reversed slope caused the kamacite crystals to end up Ni-poor near the interfaces (Fig. 3-2), instead of Ni-rich as was the case with taenite.

The development of Widmanstätten structure is summarized in Fig. 3-5. Here, in a series of Ni profiles computed at lower and lower temperatures, the processes just discussed can be seen operating and interacting with one another: both alloys change in composition, the interfaces between them migrate, and diffusion becomes progressively less able to supply Ni to crystal interiors.

THE RATE AT WHICH OCTAHEDRITES COOLED

Obviously the exact final distribution of Ni (M-shaped) across any particular taenite or taenite-plus-plessite area will have been determined by the following factors:

(1) The positions of field boundaries in the Fe–Ni phase diagram.
(2) The coefficients of diffusion of Ni in taenite and kamacite, as a function of temperature.
(3) The initial bulk Ni content of the octahedrite in question.
(4) The dimension of the system: how far from the center of the present taenite crystal did the bounding kamacite crystals nucleate?
(5) Nucleation temperature of the kamacite.
(6) The cooling history of the meteorite, since, all other things being equal, the more slowly it cooled, the more opportunity diffusion would have had to carry Ni to the interior of the taenite crystal, and so the higher the concentration of Ni at its center would finally become.

Now, knowing the Ni distribution in any given taenite crystal, **(1)** and **(2)** (from laboratory experiments), and **(3)** and **(4)** (which can be measured), it turns out to be possible to derive **(5)** and **(6)**. This is of great importance, because it tells us how rapidly the melted interior of the parent meteorite planet cooled. It is done by programming a computer to simulate the growth of Widmanstätten structure: **(1)** to **(4)** are entered as input and various values of **(5)** and **(6)** are tested until ones are found that produce a final Ni profile matching the meteorite's. The series of profiles in Fig. 3-5 were generated by such a computer program.

This has been done by J. I. Goldstein and R. E. Ogilvie, of Massachusetts Institute of Technology, and by the writer. We found that between about 600 and 400°C the octahedrites must have cooled very slowly, losing only 1 to 10 degrees of temperature in every million years.

Why so slowly? Metal is highly conductive and cools rapidly. A metal mass 1 km in diameter, orbiting in space, would cool completely in ~1,000 years. Such slow, steady cooling can only mean that the octahedrites were insulated—embedded deep inside a mass of poorly conducting material: a planet. From the cooling rates just cited we can calculate the thickness of insulating rocky material that must have covered the octahedrites and so estimate the dimension of the planet in which they evolved. Actually various octahedrites have yielded several different cooling rates, so it seems they have come from more than one planet. The cooling rates found, most of which lie between 10 and 1°C per million years, are appropriate to the centers (cores) of objects of radius 70 to 200 km. Thus small parent planets—asteroids—are indicated for the octahedrites. As already noted in Chap. 1, the asteroid belt is regarded as a likely source for the meteorites on other grounds.

OTHER MELTED AND DIFFERENTIATED METEORITE TYPES

A few pages back it was noted that melting inside a planet might yield two immiscible fluids, a dense metal-sulfur mixture that would sink to the center and form a core, and a lighter silicate magma that would float on top of it. If the iron meteorites are relics of such metallic cores, shouldn't there also be meteorites that were derived from the overlying silicate layers? In fact there are meteorites of appropriate types. These are the *achondrites* referred to in Fig. 2-2, rocks of igneous character. Most achondrites are similar to common terrestrial igneous rocks; clearly the same processes of magmatic fractionation gave rise to both. They are essentially devoid of nickel-iron metal.

There are also stony-iron meteorites (Fig. 3-6), most of which probably formed in transitional zones between nickel-iron cores and overlying achondritic layers. The "mass of iron found in Siberia by Professor Pallas" to which Chladni attached such importance (Chap. 1) was a stony-iron meteorite.

As Meunier has said, it would seem that from the irons, achondrites, and stony irons that have been collected and cataloged, we could reconstruct the melted and differentiated

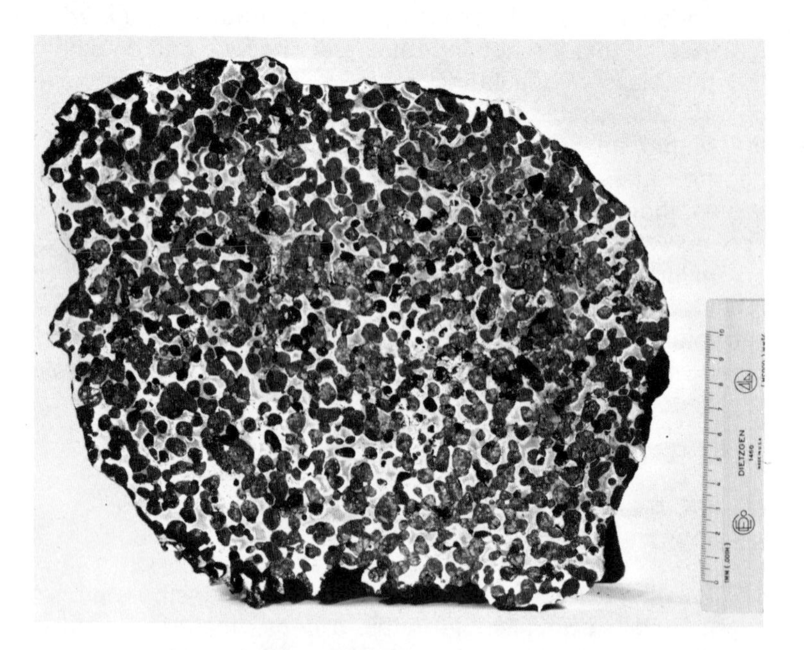

Fig. 3-6. Polished slab of Brenham, a typical stony-iron meteorite. Large, rounded crystals of olivine are densely packed into a Ni–Fe metal matrix, in which fragmentary Widmanstätten structure is occasionally visible. (*Photograph courtesy of American Museum of Natural History.*)

interior of a specimen parent meteorite planet and so learn in detail of the fractionation process. Unfortunately, whenever this is tried it quickly becomes evident that our collections are too small and incomplete. Worse, they are badly biased: some factor, perhaps a difference in the ability of various meteorite types to survive planetary breakup, or the rigors of millions of years in space or entry into the Earth's atmosphere, has caused them to come to us in nonrepresentative numbers.

EFFECTS OF PLANETARY HEATING ON THE CHONDRITES

What about the chondrites? These, too, have resided inside planets, as we shall see, but were not melted while they were there. If they had been, gravity would have drawn their metallic minerals and troilite out of them—would have separated them as it separated irons from achondrites. The fine and uniform dispersion of metal and troilite grains that we observe in chondrites (Fig. 2-1) could not have survived bulk melting.

This is not to say that the chondrites have not been heated. Most of them have been, quite severely. This can be seen from their textures. Although some chondrites show a sharply defined chondrules-in-matrix texture (Fig. 2-3), most do not. Instead their chondrules are all but lost in a relatively coarse and uniform pattern of silicate grains (Fig. 3-7): in many places the boundaries between chondrules and surrounding material have faded away altogether. Clearly these are thermally recrystallized textures, an observation first made by none other than H. C. Sorby. When polycrystalline materials are held at high temperatures for long times, their textures tend to become coarser and more homogeneous. Heat has transformed most chondrites from the Fig. 2-3 state to that seen in Fig. 3-7.

There are also important mineralogical differences between these two states of chondritic matter. They can be summed up by saying that chondrites with sharply defined

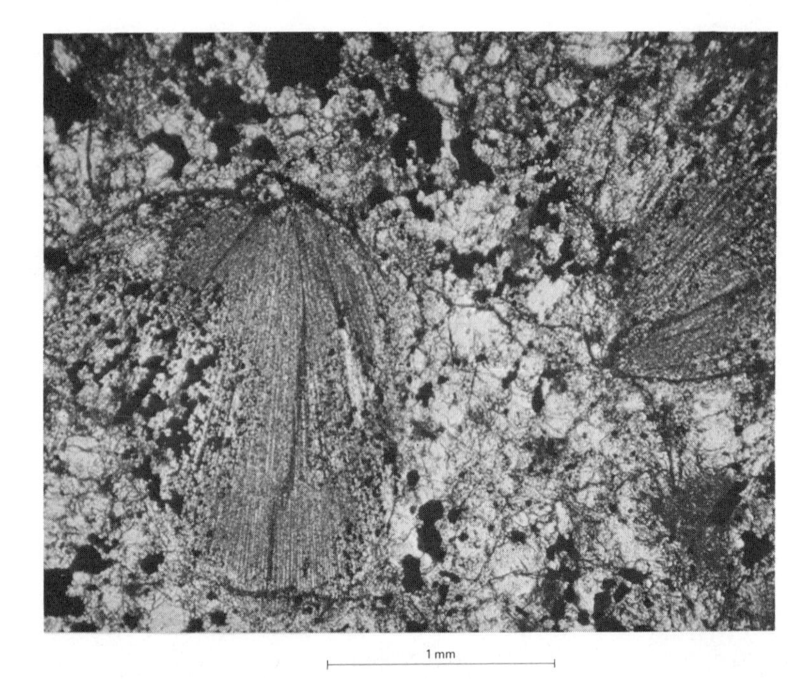

Fig. 3-7. Thin section of the Lumpkin chondrite. In this case chondrules have partly faded away into the coarsely granular silicate material between them; the chondrite has been *recrystallized.* Compare with Fig. 2-3 (note difference in scales). Most chondrites are more like Lumpkin than Tieschitz. (*Figure from J. A. Wood, Physics and Chemistry of Meteorites, "The Moon, Meteorites, and Comets," vol. IV of B. M. Middlehurst and G. P. Kuiper (eds.), "The Solar System," The University of Chicago Press, Chicago, 1963.*)

textures contain minerals that are grossly out of equilibrium with one another,* while the minerals in recrystallized chondrites have compositions that approach the equilibrium values for some temperature they have experienced. Thus

* An example of out-of-equilibrium minerals that occur in some chondrites are kamacite and pentlandite, (very approximately) $Fe_{19}Ni$ and $FeNiS_2$. Under most circumstances, these would prefer to rearrange themselves into Ni-poor sulfide (troilite, FeS) and Ni-rich metal (taenite, or taenite-plus-kamacite).

going from one state to the other has involved not only textural recrystallization but also a redistribution of chemical elements and the formation of new minerals as well as the disappearance of some old ones. Such a series of changes, occurring in terrestrial rocks at a high temperature, is termed *metamorphism*. It should be stressed that metamorphic processes operate in solid rocks: they are not effected by a melting and refreezing of silicate minerals.

How can we be sure something like Tieschitz (Fig. 2-3) evolved, by metamorphism, into Lumpkin (Fig. 3-7)? Why couldn't it have happened the other way around? Nothing is impossible, of course, but Tieschitz → Lumpkin is a vastly simpler, more straightforward, and natural process to contemplate than Lumpkin → Tieschitz. The latter would require a very complicated and improbable series of events, in order to create a mixture of nonequilibrium minerals, some low- and some high-temperature (including glass), without changing the bulk composition of the system importantly. Everything involved in going from Tieschitz to Lumpkin, on the other hand, happens spontaneously if we only postulate that the chondrite was made hot enough for a long enough time.

All different degrees of metamorphic change can be seen among the chondrites. Apparently some have been heated more severely than others. The highly metamorphosed chondrites were probably most deeply buried in the parent planets. It is convenient to distinguish six types of chondrites, on the basis of textures and mineralogy, corresponding to six levels of metamorphism. Passing over Type 1 for the moment, these are:

Type

2 Sharply defined textures; glass; nonequilibrium silicates, metal, sulfide minerals. Probably have not been metamorphosed at all.

3 Sharply defined textures; glass; nonequilibrium silicates, equilibrium metal and troilite. Somewhat metamorphosed. Tieschitz is an example.

4 Chondrule boundaries slightly blurred; glass very rare;* silicates close to equilibrium; equilibrium metal and troilite.

5 Chondrule boundaries much blurred, but many chondrules visible. No glass; no feldspar coarse enough to see in thin section. Equilibrium silicates, metal, troilite.

6 Chondrules rare, almost invisible. No glass; feldspar visible. Equilibrium silicates, metal troilite. Most highly metamorphosed. Lumpkin is an example.

Here "equilibrium" should be taken to mean that equilibrium compositions, for some particular temperature, are approached. Type 1 has been reserved for Orgueil, Ivuna, Alais, and Tonk, chondrites of a special sort: these are apparently unmetamorphosed, like Type 2 chondrites, but are intrinsically different from the latter in that they contain no chondrules.

CLASSIFICATION OF CHONDRULES

Differences of metamorphic level in chondrites have nothing to do with the *compositional* differences outlined in Fig. 2-4. Use may be made of these two independent criteria to set up a classification system for the chondrites: by ranging the six metamorphic types along one axis and the six compositional groups along another perpendicular to it, we define a grid with 36 boxes (Fig. 3-8). A chondrite's properties can then be specified closely by citing the letter and number of the box it falls in, that is, L6 or C2. Broader groupings of the boxes are often referred to by the names shown in the lower diagram of Fig. 3-8. Note that some of the 36 possible chondrite subclasses are more abundantly represented than others—in fact, about a third of the 36 have no representatives in our meteorite collections at all.

* Well-preserved chondrules often contain glass. However, sustained heating tends to destroy (devitrify) glass, transforming it into crystalline minerals, including feldspar. The feldspar in chondrites is a solid solution of $NaAlSi_3O_8$ (~80%) with $CaAl_2Si_2O_8$ (~20%).

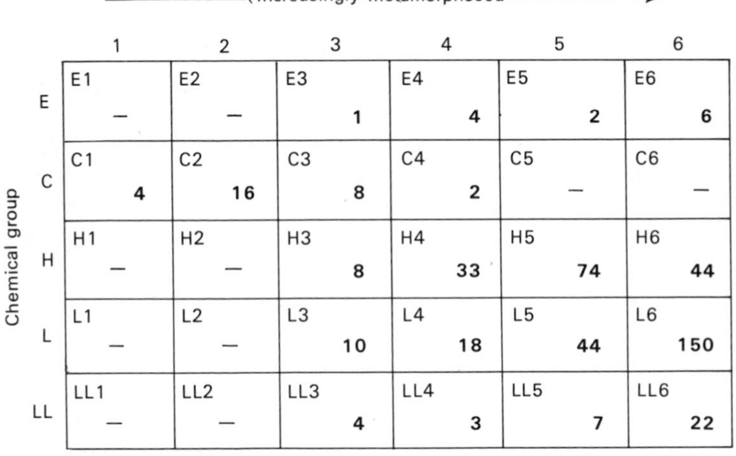

Metamorphic type (Increasingly metamorphosed →)

Chemical group	1	2	3	4	5	6
E	E1 —	E2 —	E3 **1**	E4 **4**	E5 **2**	E6 **6**
C	C1 **4**	C2 **16**	C3 **8**	C4 **2**	C5 —	C6 —
H	H1 —	H2 —	H3 **8**	H4 **33**	H5 **74**	H6 **44**
L	L1 —	L2 —	L3 **10**	L4 **18**	L5 **44**	L6 **150**
LL	LL1 —	LL2 —	LL3 **4**	LL4 **3**	LL5 **7**	LL6 **22**

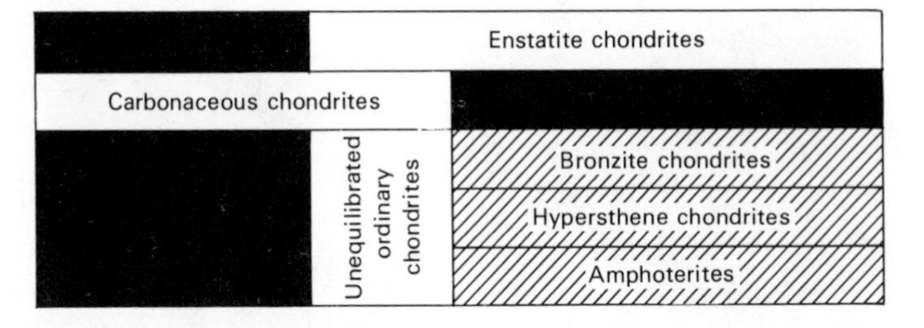

Fig. 3-8. A classification system for chondrites, based on compositional and metamorphic differences. Boldface numbers in upper diagram tell how many examples are known in each category. Chondrites may be characterized explicitly by letter-number combinations (above: L5, H4, etc.), or referred to the broader groupings in the lower diagram. Shaded block comprises the *ordinary chondrites.*

THE REDISTRIBUTION OF ELEMENTS IN CHONDRITES DURING METAMORPHISM

It is interesting to look at the way the various chemical elements are distributed inside the chondrites, on a microscopic scale. The distribution can be quite different in unmetamorphosed from what it is in metamorphosed chondrites, although both may contain essentially the same total amount of the element in question. Figures 3-9 and 3-10 show that the distribution of iron in the two cases in radically different. In Renazzo, an unmetamorphosed (C2) chondrite, iron (largely in the form of magnetite, Fe_3O_4), is concentrated in the opaque matrix material between the chondrules. It is magnetite that makes the matrix material black and opaque, in fact. Silicate minerals in the chondrules contain very little iron. In Ausson, a metamorphosed (L5) chondrite, on the other hand, iron is evenly distributed through the silicate

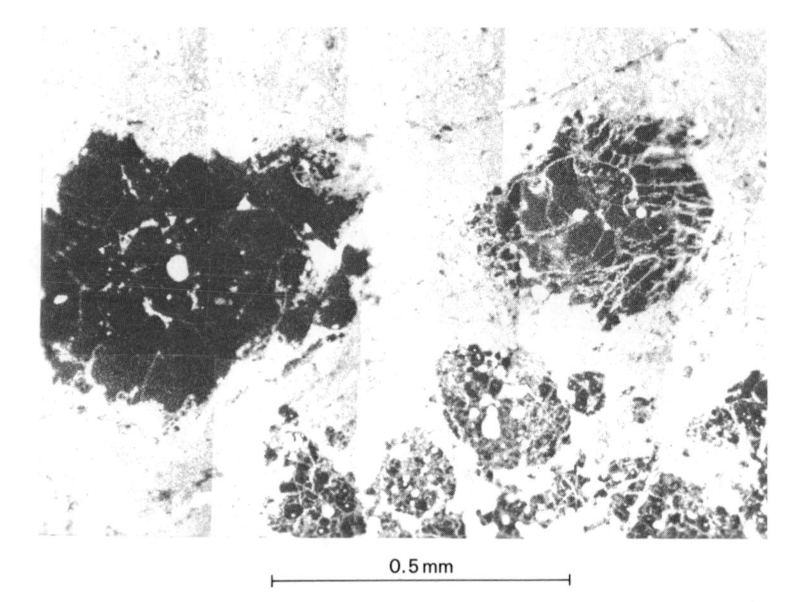

Fig. 3-9. A display of the distribution of Fe in the Renazzo chondrite, made by scanning the beam of an electron microprobe over the sample surface (in a pattern like a television raster), detecting X-radiation at the Fe k_α wavelength, and displaying the Fe pulses on a synchronously scanned oscilloscope tube. The lighter an area, the higher its Fe content. Chondrules are seen to be Fe-poor (~1 wt. %), while embedding matrix material contains ~20% Fe. Bright white spots are metal globules. (*Smithsonian Astrophysical Observatory photograph.*)

minerals (olivine and orthopyroxene), both within and between the chondrules. There is no magnetite. This is the case in all ordinary (metamorphosed) chondrites, as K. Keil and K. Fredriksson, of the University of California, have pointed out.

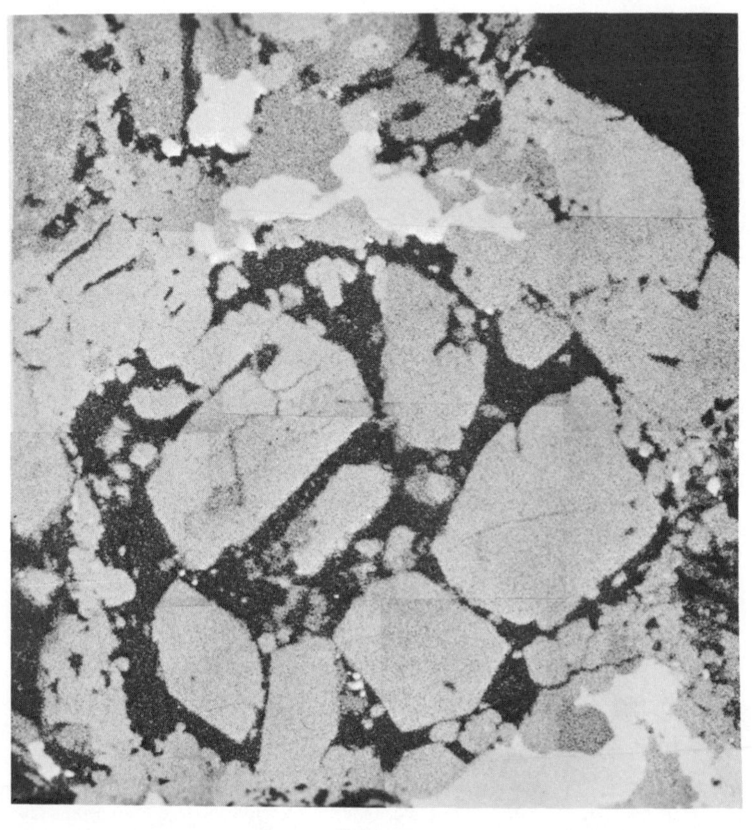

Fig. 3-10. Beam-scanning display of Fe distribution centered about a chondrule in the recrystallized Ausson chondrite. Whitest masses are metal grains. Fe is evenly distributed in the olivine (light gray, ~17%) and pyroxene (slightly darker, ~11%) crystals, inside and outside of the chondrule. Darker interstitial matter is rich in feldspar, which contains no Fe. (*Smithsonian Astrophysical Observatory photograph.*)

These cannot be intrinsically different, unrelated rocks. It would be too much of a coincidence for rocks mineralogically so different to have such similar bulk chemical compositions (including iron content) and textures (i.e., the presence of chondrules). They must be closely related: it seems clear that metamorphism must be able to turn a Renazzo-like iron distribution into an Ausson-like one.* The Renazzo distribution is a nonequilibrium one; the elements in Renazzo would prefer to rearrange themselves into the Ausson configuration, thus (approximately):

Fe_3O_4	+	Fe	+	$4MgSiO_3$	$\rightarrow$	$4FeMgSiO_4$
magnetite in matrix		in metal grains		Fe-poor pyroxene in chondrules		Fe-bearing olivine throughout the chondrite
		Renazzo				Ausson

In order for this to happen, Fe and other elements would have to be able to move around inside the chondrites. However, a sufficiently severe metamorphism would have made this possible: at high temperatures Fe and other elements can diffuse along crystal boundaries and through crystal lattices, just as Ni does through metal crystals (discussed earlier in the chapter).

THE COOLING RATE OF CHONDRITES AFTER METAMORPHISM

Diffusion coefficients in silicate minerals and rates of textural recrystallization have not been measured. We are not

* A lengthy caveat is needed here. Renazzo (C group) would not really turn into Ausson (L group) if it were metamorphosed, because there are intrinsic differences in the compositions of the C and L groups (Fig. 2-4). For this reason the comparison drawn above is somewhat misleading; it would have been better to use members of only one chemical group, comparing an L2 with an L5, or a C2 with a C5. Unfortunately there are no L2 chondrites in our collections, but this does not mean that such material never existed in the planets; by analogy with C2, it seems likely that it did. On the other hand there is a C4 chondrite (Coolidge) and probably a C5, Karoonda; Fe is distributed in the silicates of these essentially as it is in Ausson. Ausson was used in the discussion instead of Karoonda only because no electron beam-scanning pattern of the latter is available.

yet in a position to say anything definitive about the time the chondrites spent at highest temperatures, when most of their metamorphism must have been accomplished. At a guess, peak temperatures may have been in the neighborhood of 1000°C. However, chondrites of Type 3 and greater contain crystals of kamacite and taenite, and these have recorded the last, cooling phase of the chondrites' temperature histories, just as they did in the octahedrites.

Chondritic kamacite and taenite have compositions stable at 400 to 550°C, according to the Fe–Ni phase diagram. Profiles of Ni content taken across isolated taenite grains are M-shaped just like the taenite profiles in octahedrites (Fig. 3-2), showing somewhat surprisingly that taenite evolved similarly in the two cases. M-shaped taenite profiles are built up by reaction between taenite and kamacite during cooling, as we saw earlier in this chapter: this can occur naturally enough in octahedrites, where the two alloys abut against one another, but in chondrites the metal tends to occur as isolated grains of pure kamacite and pure taenite (Fig. 3-11). Evidently at 400 to 500°C Ni and Fe are able to move easily from one metal grain to another through the silicate minerals separating them, so that effectively the surfaces of taenite grains are "in contact with" surfaces of kamacite grains. Movement may be by diffusion along silicate grain boundaries, or possibly as the vapors of volatile Fe and Ni compounds.

In any case, the method of determining the cooling rates of octahedrites that was described earlier can be adapted to chondritic metal. When this is done, the chondrites are found to have cooled about as slowly as the octahedrites did, usually 1 to 10°C each million years. Thus the chondrites too must have been surrounded by insulating material—must have resided in planets. We can calculate the thickness of the insulation, i.e., how deeply the chondrites were buried in their planets as they cooled from metamorphism: some tens of kilometers deep, perhaps over 100 km. In the case of the octahedrites, these depths could be related with some confidence to the total size (radii) of the parent planets, since it seems likely that the octahedrites came from cores at the

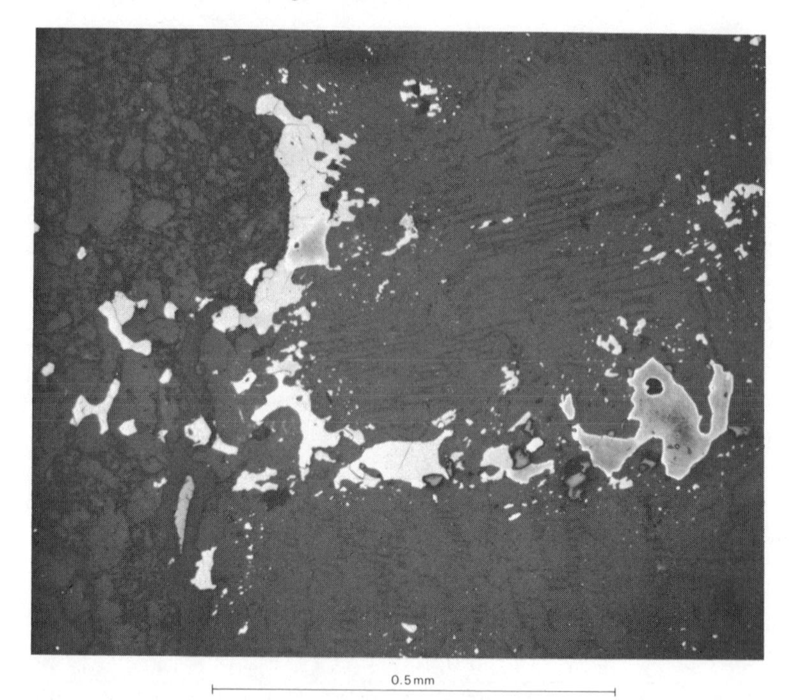

Fig. 3-11. Photomicrograph of a polished section of the Bjurböle chondrite, showing finely dispersed metal grains. Grains with gray-etched interiors are taenite, with Ni-rich rims. Plain white grains are kamacite; cracked white grains are troilite; embedding silicate material is dark gray. (*Figure from J. A. Wood, Chondrites: Their Metallic Minerals, Thermal Histories, and Parent Planets,* Icarus, *vol. 6, pp. 1–49, 1967. Courtesy of Academic Press Inc., New York.*)

centers of the planets. There is no reason to suppose that the chondrites came from the centers of planets, however; any amount of planetary material can have underlain the positions where they evolved, so nothing can be said about the total size of the parent chondrite planets.

THE SOURCE OF HEAT IN METEORITE PLANETS

An important question to ask is, what made the parent meteorite planets hot enough to metamorphose and melt? The principal known source of planetary heat is long-lived radioactivity—the decay of K^{40}, U^{238}, U^{235}, and Th^{232}. Energy released by these four nuclides as they decay in the earth is believed to be just about adequate to account for the heat that flows out of our planet. But it has become evident that K, U, Th decay was not the principal heat source in the parent meteorite planets.

In the parent octahedrite planets, it would not have been a powerful enough heat source. We have seen that these planets were small; small objects lose heat readily, so energy has to be poured into them rapidly in order to bring them up to a high temperature. If we assume that it contained as much K, U, Th as the chondritic meteorites, a planet of 200 km radius would not have heated up to more than 300°C. It would have to contain four times more long-lived radioactivity than the chondrites in order to attain the melting temperature of iron (1535°C), and this is not reasonable.

Chondrites may not have come from the same planets as octahedrites; the parent chondrite planets may have been much larger than 200 km in radius, and if so, K, U, Th decay could have heated their interiors to metamorphic temperature. However, most chondrites did not come from deep in the interiors of their parent planets: the cooling rates usually found, 1 to 10°C per million years, point to depths of 30 to 50 km. So near the surface as this, K, U, Th heating would not have raised the temperature above ~200°C, whereas we know from their metallic minerals that ordinary chondrites have been above 550°C.

In calculating the effectiveness of K, U, Th heating above, it was assumed that the planets formed and began to accumulate heat 4.5×10^9 years ago. Use of this value as the age of the solar system will be justified in the next chapter, but for the time being, what if it were wrong? Suppose the planets are much older than this? The longer ago they accreted, the more K, U, Th they must have contained at the outset in order that subsequent decay should have left the amounts of these elements we observe today. The more K, U, Th they contained, of course, the higher the initial rate of decay and heat generation and the greater the peak temperature would have been.

The chondritic metal grains refute even this possibility. If the chondrites had been heated by K, U, Th decay, they would have cooled very slowly—no faster than the rate at which these elements decayed away. The maximum possible cooling rate can be calculated from the half-life of the principal radionuclide, K^{40}, to be ~0.3°C per million years (at 500°C). As we have seen, the chondrites actually cooled much more rapidly than this.

What did heat the parent meteorite planets, then? It appears that a fierce but short-lived heat source was present when the planets formed, or shortly thereafter. The best possibility that has been raised so far is short-lived radioactivity. Perhaps when the chemical elements of our solar system were created, there were not only the stable elements and long-lived radionuclides that we observe today, but also some radionuclides with very short half-lives. These would have decayed away essentially to nothing soon after the elements were formed, of course, so they are not in the present inventory of naturally occurring terrestrial nuclides. But if the planets accreted fairly soon after the formation of the elements, some of this short-lived material would have been incorporated. Its rapid decay within the planets would have generated heat in a brief pulse, which is a far more effective way of raising temperatures in small bodies than is the gradual decay of K, U, and Th.

Actually there is evidence that two such short-lived radionuclides, Pu^{244} and I^{129}, were indeed present in the parent meteorite planets, as we shall see in the next chapter. However, these are unlikely to have heated the planets substantially. Of all the known short-lived radionuclides, the one that seems most likely to have been able to do the job is Al^{26}, as R. A. Fish, G. G. Goles, and E. Anders, of the University of Chicago, have pointed out. The half-life of Al^{26} is so short (720,000 years) that essentially all of it would have decayed and contributed its heat to a planet before any important amount of heat could leak out (by conduction), even if the planet were only a few tens of kilometers in dimension. Only ~0.2 parts per million of Al^{26} would have had to be present in a planet when it accreted in order to melt its interior.

It has not been possible to establish whether Al^{26} decay really did heat the young planets. The way to do so would be to look in the meteorites for Mg^{26}, which Al^{26} turns into when it decays. Unfortunately, meteorites, like the rest of solar system, contain a great amount of Mg^{26} (one of the common isotopes of magnesium), which was created when the elements in general were formed. The tiny amount of radiogenic Mg^{26} that meteorites might contain would be lost in this ocean of primordial Mg^{26}.

SUGGESTED FURTHER READING ON TOPICS IN CHAPTER 3

1 Iron meteorites, descriptive: S. H. Perry, The Metallography of Meteoric Iron, *U.S. Nat. Museum Bull.* 184, 1944.

2 Cooling rates of iron meteorites: J. A. Wood, The Cooling Rates and Parent Planets of Several Iron Meteorites, *Icarus,* vol. 3, pp. 429–459, 1964; and J. I. Goldstein and J. M. Short, The Iron Meteorites, Their Thermal History, and Parent Bodies," *Geochim. Cosmochim. Acta,* vol. 31, pp. 1733–1770, 1967.

3 Size of the parent planets: a few meteorites contain diamonds, and the question of whether or not these are proof of great hydrostatic pressures and therefore burial inside larger-than-asteroidal planets has been hotly debated. Arguments *pro* hydrostatic pressure are given in N. L. Carter and G. C. Kennedy, *J. Geophys. Research,* vol. 71, pp. 663–672. Arguments *con* by E. Anders, M. E. Lipschutz, D. Heymann, and B. Neilson appear in the same journal, vol. 71, pp. 619–641, 643–661, and 673–674.

4 Achondrites and stony irons, descriptive: B. Mason, "Meteorites," chaps. 7 and 8, John Wiley & Sons, Inc., New York, 1962. Other aspects of meteorites are also well treated.

5 Metamorphism of chondrites: for textural aspects, see G. P. Merrill, On Metamorphism in Meteorites, *Bull. Geol. Soc. Am.,* vol. 32, pp. 395–416, 1921. Chemical changes that have accompanied metamorphism are discussed in R. T. Dodd, W. R. Van Schmus, and D. M. Koffman, A Survey of the Unequilibrated Ordinary Chondrites, *Geochim. Cosmochim. Acta,* vol. 31, pp. 921–952, 1967. A dissenting opinion is voiced in A. M. Reid and K. Fredriksson, Chondrules and Chondrites, P. H. Abelson (ed.), "Researches in Geochemistry," vol. II, John

Wiley & Sons, Inc., New York, 1967. These authors do not believe that metamorphism has significantly affected the chondrites.

6 Cooling rates of chondrites: J. A. Wood, Chondrites: Their Metallic Minerals, Thermal Histories, and Parent Planets, *Icarus,* vol. 6, pp. 1–49, 1967.

7 Classification of chondrites: W. R. Van Schmus and J. A. Wood, A Chemical-petrologic Classification for the Chondritic Meteorites, *Geochim. Cosmochim. Acta,* vol. 31, pp. 747–766, 1967. Gives letter-number classifications of 460 chondrites.

8 Planetary heating by short-lived radionuclides: R. A. Fish, G. G. Goles, and E. Anders, The Record in the Meteorites. III. On the Development of Meteorites in Asteroidal Bodies, *Astrophys. J.,* vol. 132, pp. 243–258, 1960.

4

AGES OF THE METEORITES

> As is well known, the most exact way of determining the ages of rocks depends upon the regularity of radioactive decay processes. Obviously the same method can be applied to meteorites. . . . In our present state of ignorance of how they were formed, we must admit the possibility that there may be meteorites substantially older than the oldest strata of the earth. . . .
>
> F. A. Paneth (1928)*

Everyone is familiar with the principle of radioactive decay. Certain radioactive substances, *parent nuclides,* tend to transform spontaneously into specific *daughter nuclides.* For example:

Parent	Daughters	Half-life, years
$Rb^{87} \longrightarrow$	Sr^{87}	47×10^9
$U^{235} \longrightarrow$	$Pb^{207} + 7He^4$	0.71×10^9
$U^{238} \longrightarrow$	$Pb^{206} + 8He^4$	4.51×10^9

* From F. A. Paneth, H. Gehlen, and P. L. Guenther, Über den Helium-Gehalt und das Alter von Meteoriten, *Z. Elektrochem.,* vol. 34. pp. 645–652.

In many cases we know very precisely how rapidly the decay progresses—the *half-life* of the radioactivity expresses this. Thus, of a given amount of Rb^{87}, just half will have decayed to Sr^{87} after 47×10^9 years have elapsed. Half of the remaining Rb^{87} will decay in the next 47×10^9 years, which means that ¾ of the initial amount of Rb^{87} will have transformed after 94×10^9 years, ⅞ after 141×10^9 years, and so forth.

Every rock and meteorite contains radioactive nuclides in small amounts, and sometimes these can be used to "date" the rock or meteorite, to tell how "old" it is. In the simplest terms, what is done is to determine the concentrations of a parent and a daughter nuclide, and then make use of the half-life of the decay scheme to calculate how long it must have taken for parent to have transformed to daughter to that extent.

In practice it is, of course, a much more complicated and difficult business. But this has not stopped a number of dedicated workers in European and American laboratories from establishing the ages—as we shall see, there are many different kinds of "ages"—of a very large number of meteorites. Their work is of capital importance: only through radioactivity have we any hope of knowing the time scale of the early events that have affected the meteorites and, presumably, the planets in general.

Rb^{87}–Sr^{87} AGES

The Rb^{87}–Sr^{87} decay scheme noted above has been used to "date" some meteorites: let us discuss it first and in some detail. Two important points were glossed over in the simple-minded account of radioactive dating just given. (1) What do we mean by the "age" of a meteorite, anyway—the length of time since its parent planet accreted, or the length of time since the nuclides of which it was composed were formed, or just what? (2) What if all the Sr^{87} we find in a meteorite is not the daughter of Rb^{87} decay? Suppose some Sr^{87} happened to be already present at the outset (whatever this may mean)? How can we tell the radiogenic Sr^{87} from this primordial Sr^{87}, and so calculate an honest "age" for the meteorite?

One way of getting around (2) is to consider not a single meteorite, but a group of them, and to postulate that at one time all contained Sr of the same isotopic makeup: the ratio Sr^{87}/Sr^{86} was the same for all. (Sr^{87} is a radiogenic daughter, remember; Sr^{86} is not: nothing creates or destroys Sr^{86} to an important extent.) Call this time of uniform Sr^{87}/Sr^{86} the "beginning," for the present. Now, if the ratio of total Rb to total Sr present varies from one meteorite to another in the group defined, (2) can be gotten around. The higher Rb/Sr (or, to be more precise, Rb^{87}/Sr^{86}) is, the faster Rb^{87} decay will have added Sr^{87} to the total Sr present and the more the ratio Sr^{87}/Sr^{86} will have increased in the time that has elapsed since the "beginning." The ratio Sr^{87}/Sr^{86} will have gotten increasingly dissimilar among the meteorites with the passage of time.

The effect is shown graphically in Fig. 4-1, a plot of Sr^{87}/Sr^{86} against Rb^{87}/Sr^{86}. At the "beginning" postulated, Sr^{87}/Sr^{86} was the same in all meteorites, so points representing them at that time plot along a horizontal line. With the passage of time, however, the line rises and tilts: its left end remains fixed, of course, because there $Rb^{87} = 0$, so no decay occurs and Sr^{87} cannot increase.

Points representing 13 meteorites are entered in Fig. 4-1. P. W. Gast, of Columbia University, and W. H. Pinson and coworkers at Massachusetts Institute of Technology obtained the Sr^{87}/Sr^{86} and Rb^{87}/Sr^{87} data for these by chemically extracting and separating the Sr and Rb they contained, then analyzing the Sr and Rb isotopic compositions mass-spectrometrically (Fig. 4-2). The 14 points define an almost-straight, sloping line. The fact that the line is straight vindicates our assumption that there was a "beginning" when Sr everywhere had the same isotopic makeup: there is no other initial state of affairs that can plausibly be postulated which subsequent Rb^{87} decay could turn into the observed linear array of points.

From the half-life of Rb^{87}, it can be calculated that about

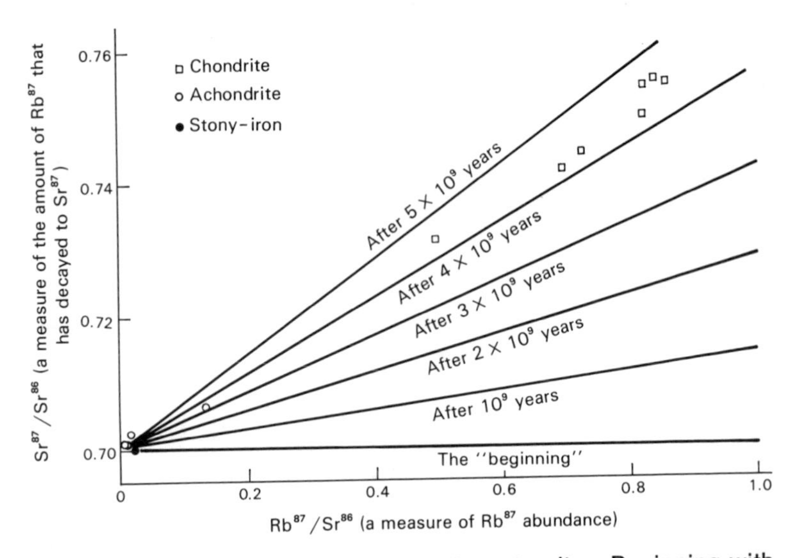

Fig. 4-1. Sr^{87} and Rb^{87} abundances in meteorites. Beginning with a situation where Sr^{87}/Sr^{86} was everywhere equal to 0.70, lines show how radioactive decay of Rb^{87} would have changed relative abundances with time. Measurements actually made in meteorites (points) show that $\sim 4.4 \times 10^9$ years have elapsed since meteorites had uniform Sr^{87}/Sr^{86}.

4.4×10^9 years of decay was required to bring the line of meteorite points from a horizontal position up to its present slope. But just what was it that happened 4.4×10^9 years ago, at the "beginning"? What does a Rb^{87}–Sr^{87} age mean? Apparently the material of the parent meteorite planets existed in a homogeneous, well-mixed state (or series of states) prior to the "beginning," so that not only Sr^{87}/Sr^{86} but also Rb^{87}/Sr^{86} was everywhere the same. If Rb^{87}/Sr^{86} had been inhomogeneous, decay would have made Sr^{87}/Sr^{86} inhomogeneous, too. After the "beginning," the meteorites had different Rb^{87}/Sr^{86} ratios, as we have seen. The "beginning," then, was an event that chemically fractionated

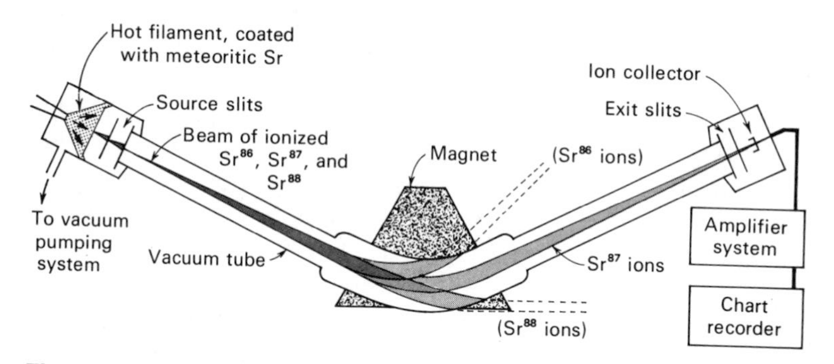

Fig. 4-2. Mass spectrometer (simplified). Filament (left) vaporizes and ionizes Sr. Potential difference of several thousand volts between filament and slit plates accelerates ions toward center of system (magnet). Slits collimate ions into a narrow beam. The magnetic field bends the path of each ion in the beam by an amount dependent on its mass; in the example sketched above, the magnetic field strength is such that Sr^{87} ions are deflected in their paths just enough to continue down the spectrometer tube to the ion collector. Sr^{86} ions are deflected too much, Sr^{88} ions too little; both are absorbed by the walls of the spectrometer tube. By varying the magnetic field strength, however, the Sr^{86} or Sr^{88} ion beam can be brought to focus on the ion collector instead of Sr^{87}. The ion collector measures the beam intensity; by surveying the intensities of all Sr isotope beams, the isotopic composition of Sr placed on the filament can be determined with great accuracy.

the meteorites, giving more Rb to some of them than to others.

The six points farthest left in Fig. 4-1 are achondrites and a stony-iron (magmatically fractionated meteorites), while the points toward the right are chondrites (which have a generalized composition and have not been bulk melted). Magmatic fractionation is quite capable of yielding rock types with low Rb and high Sr content, as is the case with the achondrites. Evidently, this is what the Rb^{87}–Sr^{87} age means, at least so far as these two clusters of points are concerned: melting occurred in the parent plants some 4.4×10^9 years ago, and achondritic layers formed in which the Rb^{87}/Sr^{87} ratio was lower than the chondritic value.

One other chondrite, Beardsley, falls far outside Fig. 4-1, having a Rb^{87}/Sr^{87} ratio four times higher than the other chondrites; but its Sr^{87}/Sr^{86} ratio is correspondingly higher, so that an age of about 4.4×10^9 years is still indicated. It is not clear what caused Rb to become concentrated in Beardsley relative to the other chondrites $\sim 4.4 \times 10^9$ years ago. All the chondrites in Fig. 4-1 are metamorphosed, and Rb (which is quite volatile) may have entered Beardsley during the metamorphic event. Or Rb may have been fractionated at the time the chondrites formed (a process we have not yet discussed).

URANIUM-LEAD AGES

The uranium-to-lead decay schemes noted at the beginning of this chapter form the basis of another method of dating meteorites—one which has much in common with the Rb^{87}–Sr^{87} method. U^{235} and U^{238} have different half-lives, so their daughters Pb^{207} and Pb^{206} are produced at different rates, making the ratio Pb^{207}/Pb^{206} in meteorites change with time. As was the case with Rb^{87}–Sr^{87}, Pb^{207}–Pb^{206} ages have been determined only for pairs or groups of meteorites, not for single meteorites, and the event dated was again an era of chemical fractionation, this time of Pb from U.

The first Pb^{207}–Pb^{206} age of meteorites was gotten by C. C. Patterson, of the California Institute of Technology. By now data for a dozen or more meteorites are at hand, and none are badly out of line with an age of 4.5×10^9 years. Meteorites studied include chondrites, achondrites, and irons, and it seems clear that the event dated was the same melting and magmatic fractionation that the Rb^{87}–Sr^{87} age refers to. The two ages cited are not known precisely enough for any importance to be attached to the small difference between them (10^8 years).

GAS-RETENTION AGES

Some decay schemes yield gaseous daughter nuclides:

Parent		Daughters	Half-life, years
K^{40}	(11%) →	A^{40}	
	→	Ca^{40}	1.25×10^9
U^{235}	→	$Pb^{207} + 7He^4$	0.71×10^9
U^{238}	→	$Pb^{206} + 8He^4$	4.51×10^9
Th^{232}	→	$Pb^{208} + 6He^4$	13.9×10^9

These have been used extensively to date meteorites. It might seem impractical: If the daughters are gases, wouldn't they tend to leak out of the meteorites as fast as decay produced them? Not necessarily. The parent nuclides occur in crystalline minerals in the meteorites, and so when an atom of K^{40}, U, or Th decays, the gaseous daughter nuclide finds itself trapped inside a crystal lattice. In order to escape from the meteorite, it first has to make its way out of the crystal by lattice diffusion. Lattice diffusion is highly temperature-dependent, as we saw in Chap. 3. It appears that much above about 250°C, A^{40} tends to escape from meteoritic minerals, while much below that temperature it remains trapped. He^4 atoms are smaller than A^{40} atoms and slip more readily through crystal lattices, so the critical temperature separating He^4 loss from retention is lower, about 100°C. (These temperatures are *very approximate.*)

The K^{40}–A^{40} and U,Th–He^4 techniques gives us *gas-retention ages,* then—they date the length of time that has elapsed since a meteorite last cooled beneath the critical temperatures at which A^{40} and He^4 can begin to accumulate. The concept is simpler than the meaning attached to Rb^{87}–Sr^{87} and Pb^{207}–Pb^{206} ages. There is no need to assume that a group of meteorites was affected by the same initial event; a gas-retention age can be gotten for each meteorite.

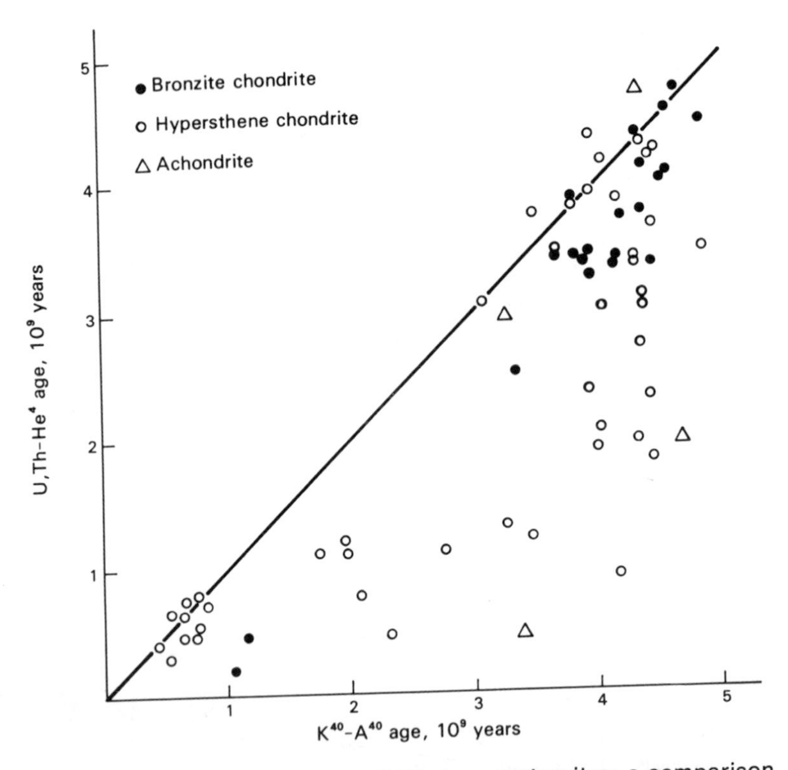

Fig. 4-3. Gas-retention ages of 69 stony meteorites: a comparison of results obtained by the U,Th–He^4 and K^{40}–A^{40} methods.

K^{40}–A^{40} and U,Th–He^4 ages for 69 stony meteorites appear in Fig. 4-3. Most of these were determined by European research groups, notably groups working at the University of Bern (Switzerland) and in the German Max Planck Institutes at Heidelberg and Mainz. Three categories of points can be distinguished in the plot:

(1) About 24 meteorites (upper right in plot) are quite old, 3.5 to 4.5 $\times$ 10^9 years, and approximately the same age is yielded by both K^{40}–A^{40} and U,Th–He^4 techniques. How well does this tie in with the Rb^{87}–Sr^{87} and Pb^{207}–Pb^{206} ages just discussed? Quite satisfactorily, it would seem.* The Rb^{87}–Sr^{87} and Pb^{207}–

Pb^{206} results speak of a severe heating event about 4.5×10^9 years ago that melted some planetary material. The gas-retention ages say that within times of 10^9 years or less after this heating event the 24 meteorites under discussion had cooled by (very roughly) 1000°C, to temperatures at which A^{40} and He^4 began to be retained. Thus cooling rates of the order of 1°C per million years or greater were necessary; but this is precisely the range of cooling rates that was indicated for ordinary chondrites by the Ni distributions in their taenite grains (Chap. 3).

(2) Ten hypersthene chondrites (lower left in the plot) also have approximately equal K^{40}–A^{40} and U,Th–He^4 ages, but the ages are very short ($\sim 0.6 \times 10^9$ years). E. Anders has argued persuasively that their parent planet suffered a catastrophic collision in space 0.6×10^9 years ago which so severely heated these chondrites that their accumulated A^{40} and He^4 was completely driven off, resetting the radioactive clock, so to speak. (Laboratory experiments have shown that compression of rocky material, as by hypervelocity impacts, can raise its temperature by many hundreds of degrees. It also produces certain physical changes in the rock—brecciation, blackening, veining, and transformations in metal grains—which are characteristic of the 10 chondrites in this category.)

(3) Finally, about 30 meteorites, mostly hypersthene chondrites, lie well below the 45° line in Fig. 4-3 and thus have U,Th–He^4 ages that are shorter than their K^{40}–A^{40} ages. Evidently these have been reheated to a degree that caused some of the accumulated gas

* However, the comparison may be irrelevant. The Rb^{87}–Sr^{87} and Pb^{207}–Pb^{206} ages are attached primarily to achondrites and irons, fractionated materials; they probably have no real meaning for chondrites, which seem not to have been fractionated. Gas-retention ages, on the other hand, have been gotten for chondrites and little else (Fig. 4–3). Thus the above cooling-rate calculation is made on the basis of when the nonchondrites heated up and when the chondrites cooled off: it is assumed that the same heating event affected both types of meteorite. This cannot be defended rigorously.

to be lost, but not all of it. Under these circumstances the gas-retention "ages" have little meaning in terms of a time interval. He^4 losses were naturally greater than those of A^{40}, because He^4 diffuses through crystals more readily, so the apparent He^4 ages are always shorter than the A^{40} ages. Several possible heat sources may have caused the gas losses: the most likely are collisional reheating and orbits that carried the meteorites close to the sun.

COSMIC-RAY EXPOSURE AGES

Everything in space is constantly being bombarded by cosmic rays (mostly very-high-energy protons). When these strike the nuclei of atoms, they tend to chip or spall some of the protons and neutrons out of them. This changes the identity of the struck atoms, of course: take one proton away from an Fe^{56} nucleus, and it becomes Mn^{55}. A vast number of spallation reactions are possible, but a typical one might look like this:

$$Fe^{56} + H^1 \rightarrow Cl^{36} + \underbrace{H^3 + 2He^4 + He^3 + 3H^1 + 4 \text{ neutrons}}$$

Fe^{56}	H^1	Cl^{36}	$H^3 + 2He^4 + He^3 + 3H^1 + 4$ neutrons
nucleus before being struck	high-energy proton	remainder of Fe^{56} nucleus	debris spalled off from Fe^{56} nucleus (plus impacting proton)

Cosmic rays are largely absorbed by a meter or so of rocky material, so the meteorites were shielded from cosmic radiation as long as they were inside planets or sizable fragments of planets. From the time they were broken down to pieces less than a few meters in dimension until they were captured by the Earth, however, they were exposed to cosmic radiation. Reactions like the above, between cosmic rays and all the types of atoms that go to make up meteorites, generated a whole spectrum of new "cosmogenic" nuclides. Many of these were nuclides already present in abundance in the meteorites, but others were nuclides that would otherwise be rare or absent (for example, H^3, He^3, Ne^{20}, Ne^{21}, Ne^{22}, A^{36}, and A^{38}). By measuring the amounts of such identifiable cosmogenic nuclides present in a meteorite, it is possible to determine how long the latter was exposed to cosmic rays in space, since of course the longer it was exposed the more cosmogenic nuclides it accumulated.

The best procedure is to measure the abundances of two cosmogenic nuclides in a meteorite, one stable and one radioactive, such as the pair A^{38}–A^{39}. Spallation reactions generate A^{39} (the radioactive nuclide) at a steady rate while a meteorite orbits in space; but of course the more A^{39} accumulates, the more decays each second. Eventually, after a few half-lives of A^{39}, a point is reached—a "steady state"—where A^{39} is decaying just as fast as it is being generated. The A^{39} concentration cannot increase beyond this level (lower curve in Fig. 4-4). Because the half-life of A^{39} (325 years) is so short compared to the general time scale of

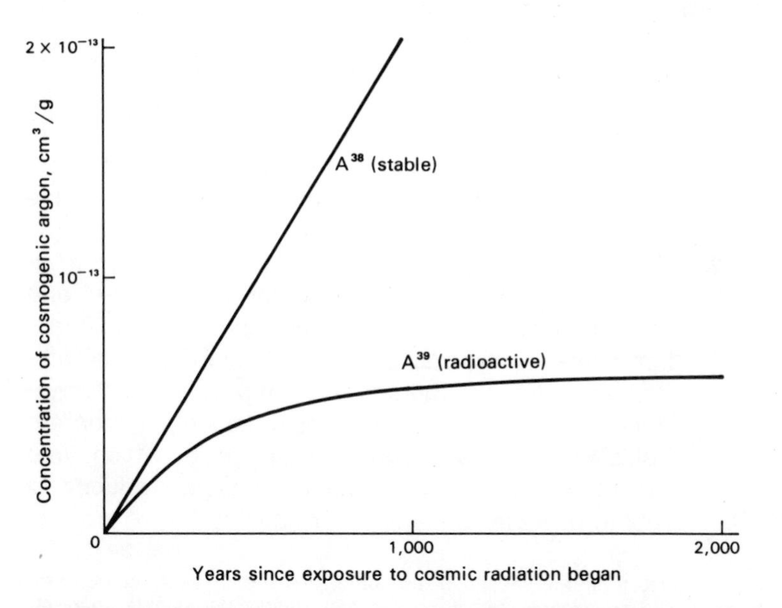

Fig. 4-4. Generation of argon isotopes by cosmic radiation, in a typical chondrite.

events that have affected the meteorites, it can be assumed that this A^{39} steady state had already been reached by meteorites by the time they fell to Earth.

The reason we want to learn the steady-state level of A^{39} in any particular meteorite, and from it the rate at which A^{39} is decaying, is that the latter is identical to the rate at which cosmic rays were generating A^{39} in that particular meteorite.* This generation rate of A^{39}, together with experimental information on the relative rates of production of A^{39} and A^{38} in targets exposed to cyclotron beams, allows us to calculate the rate at which cosmic radiation must have been producing A^{38} in the meteorite of interest. From this, finally, and the amount of A^{38} that has actually accumulated in the meteorite, the latter's *cosmic-ray exposure age* can be calculated.

Many meteorites have been so dated, at a number of institutions, including especially the Brookhaven National Laboratory, the University of Minnesota, and the European laboratories already noted. From the assembled data (Fig. 4-5) several interesting facts emerge:

(1) Cosmic-ray exposure ages are very much shorter than the differentiation and gas-retention ages discussed earlier. The meteorites have spent most of their lives (after cooling, but before exposure to cosmic radiation) inside cold, inert planets or sizable fragments of planets.

(2) The exposure ages are not scattered at random, but tend to cluster about certain preferred values—5 and 22 million years for bronzite chondrites, 7 and 20 million years for hypersthene chondrites, and ~700 million years for iron meteorites. Apparently, as E. Anders has stressed, each cluster represents a separate collision that has occurred in space, at which time planetary matter of a particular type was reduced to rubble of meter dimension or smaller.

* This varies from one specimen to another, depending as it does on factors like the chemical composition and how deeply the sample analyzed was buried inside the meteorite (and so partly shielded from cosmic radiation) while it orbited in space.

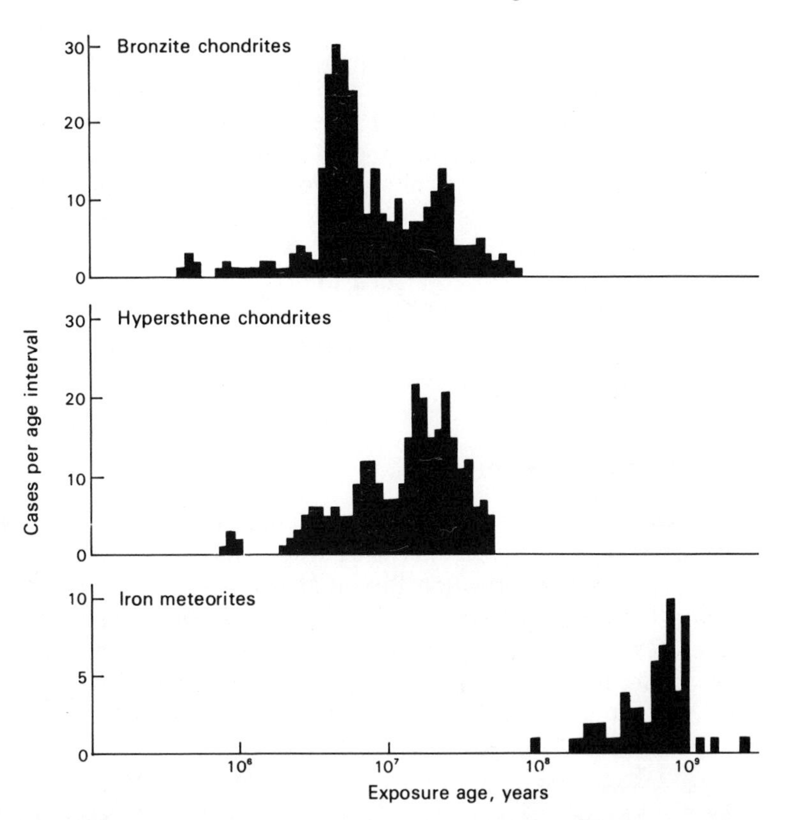

Fig. 4-5. Distribution of cosmic-ray exposure ages found in several meteorite classes. Chondrite ages are based on levels of cosmogenic Ne^{21}; iron ages, on cosmogenic isotopes of potassium (measurements by H. Voshage).

(3) Chondrites (and achondrites too, although they are not shown) have very much shorter exposure ages than do irons. This difference suggested to H. C. Urey that the two classes of meteorites must come from different parts of the solar system—the chondrites from the moon, irons from the asteroids—as has already been discussed in Chap. 1. Certainly it is true that debris from the moon would be swept up by the Earth faster than asteroidal debris, which has to wait for Mars to perturb it in just the right way to send it into the vicinity of Earth. However, the fact that meteorite orbits seem to show an affinity for the asteroid belt (Fig. 1-2) makes it appear questionable that this is the explanation for the difference in exposure ages. The alternative is that all meteorites come from the asteroids, and that the longer ages of irons reflect their greater strength—they are more likely to survive collisions in space than are stones. Objections can and have been raised to both lunar and asteroidal origins for the chondrites. The question remains unsettled.

EXTINCT RADIOACTIVITY

In Chap. 3, the possibility was entertained that some radionuclides with very short half-lives might have been incorporated in the planets when they first accreted. We would not find them in the Earth or meteorites now, of course, because they are "extinct," having decayed down to undetectable concentrations long ago. A number of possible short-lived radionuclides, potentially the most important ones, are listed in Table 4-1.

Table 4-1

Important short-lived radionuclides

Nuclide	Decay half-life, million of years
Be^{10}	2.5
Al^{26}	0.72
Cl^{36}	0.3
Fe^{60}	~0.3
I^{129}	16
Np^{237}	2.2
Pu^{244}	76
Cm^{247}	>40

Of these, I^{129} would probably have left the most conspicuous evidence of itself if it were present. I^{129} decays to Xe^{129}, and since I is inherently far more abundant than Xe in mete-

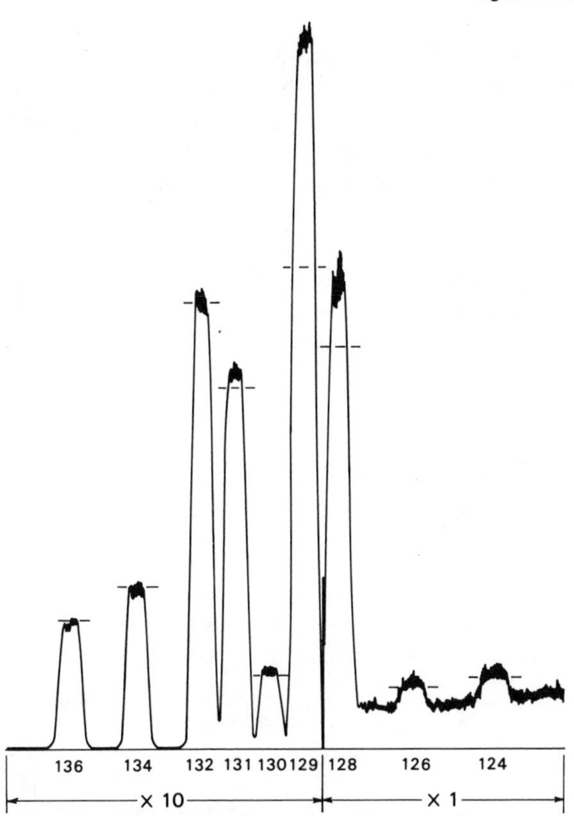

Fig. 4-6. Mass spectrum of Xe in the Richardton chondrite. Short horizontal lines show the normal isotopic composition of atmospheric Xe (relative to Xe^{132}); Richardton Xe^{129} is half again as abundant as terrestrial Xe^{129}. (*Figure from J. H. Reynolds, Determination of the Age of the Elements,* Phys. Rev. Letters, *vol. 4, pp. 8–10, 1960.*)

orites, radiogenic Xe^{129} might very well have been created in amounts large enough to alter the isotopic composition of a meteorite's Xe perceptibly. After Harrison Brown suggested the possibility of extinct radioactivity in 1947, several unsuccessful attempts were made to find excess Xe^{129} in meteorites. Not until 1960 was it detected, by J. H. Reynolds, in the Richardton chondrite. The Richardton mass spectrum (Fig. 4-6) shows 50% again as much Xe^{129} as Xe from the

Earth's atmosphere. (Why compare with terrestrial Xe? Maybe the Earth, not Richardton, is anomalous? This is improbable. It is easy to see how a positive Xe^{129} anomaly can be created by I^{129} decay, but there is no process that could *remove* Xe^{129} selectively and so create a negative anomaly in terrestrial Xe. The Xe^{129} content of "primordial" Xe in carbonaceous chondrites is essentially the same as that of terrestrial Xe: apparently this is a good approximation to the overall Xe^{129} content of solar-system Xe.) Excess Xe^{129} has since been found in a number of other meteorites, including one iron (Sardis).

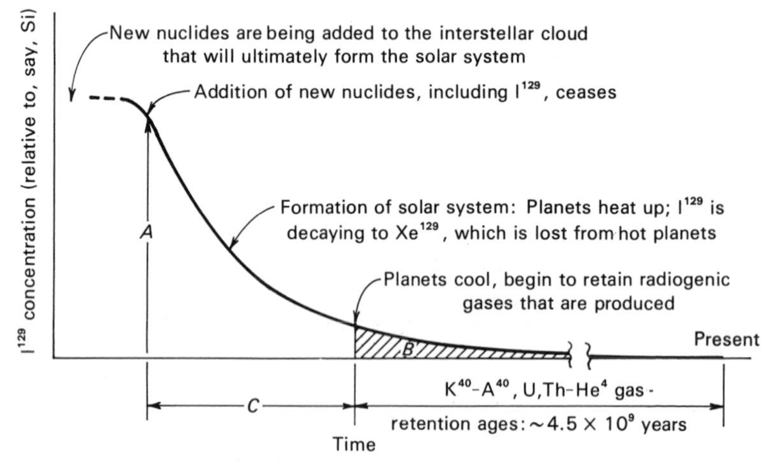

Fig. 4-7. Curve of I^{129} decay in the solar system. Given *A* and *B* (the amount of radiogenic Xe^{129} that remains trapped in a planet; this can be measured in meteorites), it is possible to calculate *C* (the I^{129}–Xe^{129} formation interval).

Xe^{129} allows another type of age, the *formation interval,* to be measured. Figure 4-7 shows the meaning of the formation interval: given *A, B,* and the half-life of I^{129}, the formation interval *C* can be calculated. *A* is not really known, but has to be estimated from theories of the origin of elements; fortunately, the formation interval is not very sensitive to errors in *A.* Formation intervals ranging from 90 to 250 million years have been gotten for the various meteorites studied so far. The differences probably reflect (among other things) unlike thermal histories experienced by these meteorites: some regions in the parent meteorite planets must have cooled to Xe^{129}-retention temperatures sooner than others.

Note that formation intervals are complementary to the gas-retention ages discussed earlier. K^{40}–A^{40} and U,Th–He^{4} ages indicated that the parent meteorite planets cooled $\sim 4.5 \times 10^{9}$ years ago, but said nothing of what went before—the solar system might already have been very old by then. Now this possibility is ruled out by I^{129}–Xe^{129} formation intervals: no more than ~200 million years of solar-system history, and probably less than that, preceded the cooling of the parent meteorite planets.

Two independent lines of research have pointed to the existence in the parent planets of another short-lived radioactivity. Uranium has several ways of spontaneously breaking down: besides decaying into Pb and He, as was discussed earlier, it tends to fission into an array of lighter nuclides, one of which is Xe^{136}. Fission releases a great deal of energy, as is well known, and this has the effect of shooting the newly produced nuclides into whatever crystal lattices happen to be in the way. There the fission "bullets" do damage, leaving a trail of displaced atoms which can persist for billions of years if the crystal is not heated (annealed). R. L. Fleischer, P. B. Price, and R. M. Walker, working at the General Electric Company (Schenectady, New York) discovered that when fission-damaged crystals are put in appropriate acid or alkaline solutions, the liquid tends to seep into the fission tracks, etch them out, and enlarge them until they are microscopically visible (Fig. 4-8). Then the fission tracks can actually be counted and the amount of fission that has occurred can be learned.

The General Electric group learned that feldspar crystals in Moore County (achondrite) and pyroxene inclusions in Toluca (iron) contain far too many fission tracks to be accounted for by the amount of U these crystals contained. At about the same time, M. W. Rowe and P. K. Kuroda at the University of Arkansas reported that the Pasamonte achondrite contains too much Xe^{136} to have been produced by that

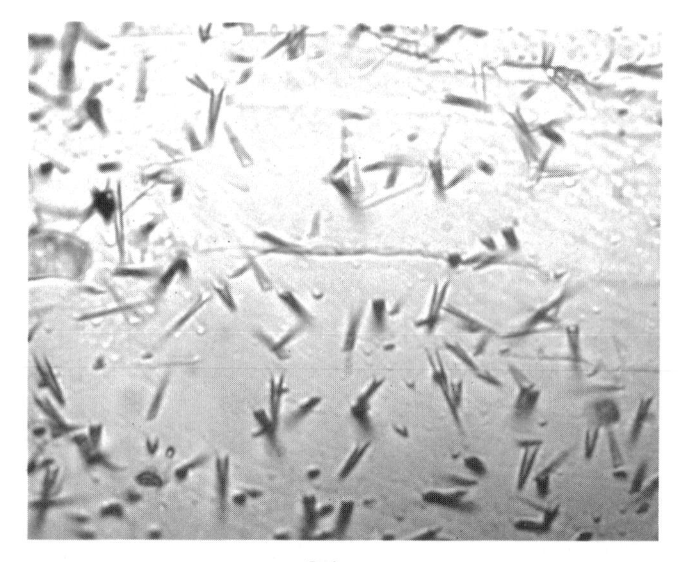

Fig. 4-8. Charged particle tracks, apparently due to fission of Pu^{244}, in a pyroxene crystal inclusion from the Odessa iron meteorite. (*Figure from R. L. Fleischer, P. B. Price, and R. M. Walker, Tracks of Charged Particles in Solids,* Science, *vol. 149, no. 3682, pp. 383–393, 1965. Copyright 1965 by the American Association for the Advancement of Science.*)

stone's concentration of U. Both groups concluded that another fissionable nuclide, Pu^{244} (decay half-life, 76 million years), must have been present when the meteorites formed. Formation intervals can be calculated for Pu^{244} just as they are for I^{129}, and the values gotten (Moore County, ~50 million years; Toluca, ~200 million years; Pasamonte, ~300 million years) are similar to I^{129}–Xe^{129} formation intervals.*

* Meteorites are considerably more complicated and their interpretation less straightforward than there is space to indicate in this book. As an isolated example of the perversity of meteorites, let us consider the two achondrites La Fayette and Nakhla. They contain anomalous amounts of Xe^{129} but not of Xe^{136}: evidence of short lived I^{129}, but not of Pu^{244}. Why? If their parent planet had cooled so slowly as to prevent accumulation of Pu^{244} fission products, then all Xe^{129} should have been excluded as well, since the half-life of I^{129} is shorter than that of Pu^{244}.

PRIMORDIAL NOBLE GASES

Many of the studies described in this chapter have been of isotopes of He, A, and Xe—the noble gases. Noble gases (including Ne and Kr) have been the favorite target of isotope studies for good reasons: they are not chemically combined with the other elements that make up a meteorite but only dissolved among them, so they can be cleanly and easily separated from the rest of the meteorite (Fig. 4-9); and, because they are not condensable at reasonable solar-system temperatures, they were largely excluded from the planets when they accreted: the scarcity of noble gases in planetary mate-

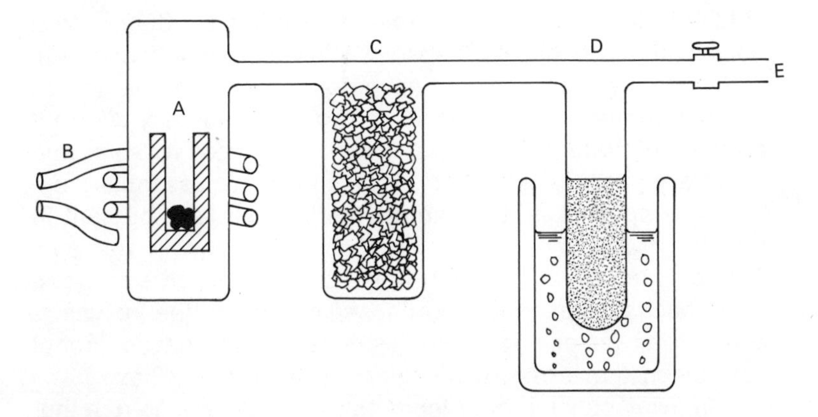

Fig. 4-9. Vacuum line for extraction and purification of meteoritic gases (simplified). The meteorite is placed in a Mo crucible (*A*) in a vacuum and melted by an RF induction heating coil (*B*) that surrounds it (outside of the vacuum line). All gases are driven off. Reactive gases such as O_2 and N_2 (atmospheric contaminants) form solid compounds when they come in contact with heated Ti foil (*C*), leaving only noble gases. Xe, Kr, and A are frozen out on activated charcoal (*D*) held at liquid N_2 temperature (−210°C); the remaining He and Ne are then admitted to the mass spectrometer (*E*) for analysis. Later when (*D*) is warmed to dry ice temperature (−78°C), argon is evolved and can be analyzed separately. Finally, Kr and Xe are removed from the charcoal for analysis by heating to 100°C.

rial (including meteorites) makes noble-gas isotopes generated by later processes (radioactive decay, fission, and cosmic-ray spallation) all the more conspicuous.

However, apparently "primordial" noble gases were not completely excluded from the parent meteorite planets, and in a few of the meteorites they are quite abundant. This was first noted in 1956 by E. K. Gerling and L. K. Levskii, of the U.S.S.R., who found that the Pesyanoe achondrite contains very large amounts of noble gases.* These could not be accounted for as radiogenic or cosmogenic products or terrestrial contamination: they must be "primordial" gases, introduced into the meteorite as gases at some very early time. A great deal of detailed research has been done on primordial noble gases in meteorites by now, but the results obtained are not yet satisfactorily understood. Space does not permit more than a glimpse of this very complex problem.

P. Signer and H. E. Suess have pointed out that meteorites contain primordial noble gases in two fundamentally different patterns of abundance. They referred to these as solar and planetary gases (Fig. 4-10). *Solar* gases are essentially unfractionated; that is, the various noble gases are present in relative proportions similar to those observed in the sun. The few meteorites in which they occur are of diverse types and most have been metamorphosed or melted (the Pesyanoe achondrite, for example), so it seems that the gases cannot have been there from the beginning: they would have been driven away by such high temperatures. Solar-gas-rich meteorites are usually breccias (assemblages of broken fragments, more or less welded together), and the research group at the Max Planck Institute for Chemistry, Mainz, has been able to show that the noble gases are not evenly dispersed through these meteorites but are concentrated near the surfaces of the breccia fragments. One possible interpretation has been suggested by H. Wänke: perhaps the fragments were once loose, impact-produced debris on the surface of a planet, such as the moon; there they were bombarded by the solar wind (the low-energy flux of particles of

* About 0.02 cm^3 of gas (largely He^4), at room temperature and 1 atm pressure, per cubic centimeter of meteorite.

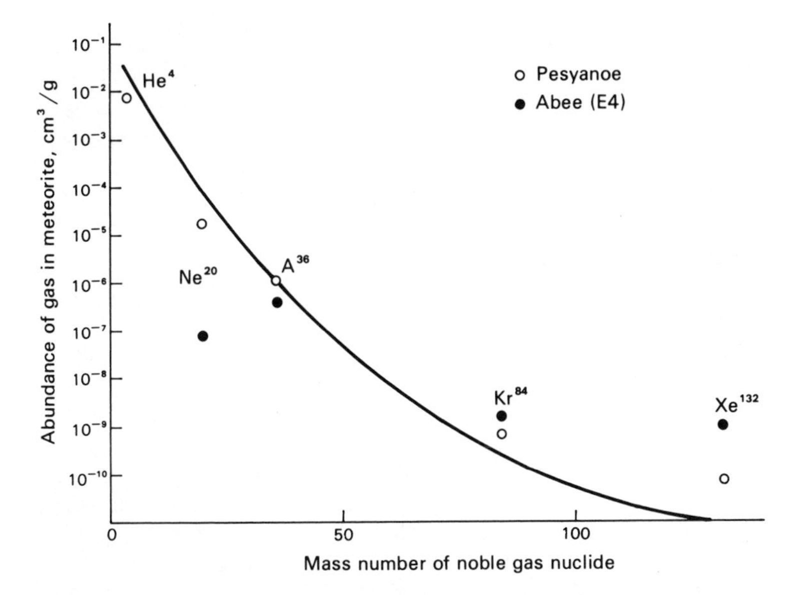

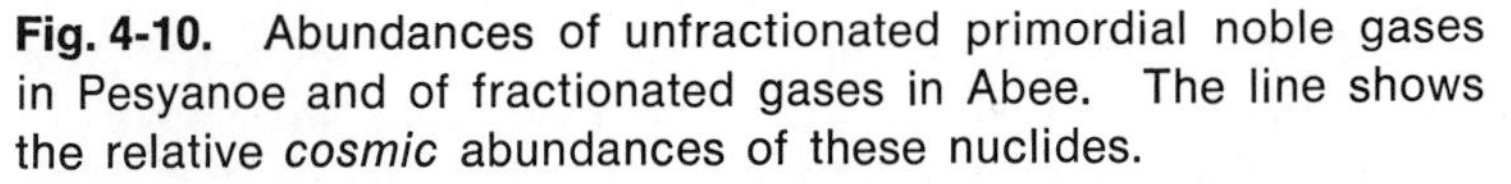

Fig. 4-10. Abundances of unfractionated primordial noble gases in Pesyanoe and of fractionated gases in Abee. The line shows the relative *cosmic* abundances of these nuclides.

all types, including noble-gas ions, that perpetually streams out of the sun), which drove noble-gas atoms into the surfaces of the fragments. Later the debris was covered over and consolidated into dense planetary material.

Planetary gases have been fractionated, in such a way that the lighter gases are depleted (Fig. 4-10). They are abundant only in relatively unmetamorphosed chondrites, which may be material that has survived essentially unchanged since the planets were formed (Chap. 5), so it is possible that these are truly primordial gases. J. Zähringer has suggested that the fractionation may have been brought about by partial diffusive losses of the noble gases from mineral crystals. Assuming that solar proportions of noble gases were once dissolved in the minerals, and that the chondrites have been mildly heated at some stage, it is easy to see that the lighter gas atoms (which are smaller in size) would tend to diffuse more easily out of crystal lattices and be lost preferentially. Zähringer has shown that the relative abundances and isotopic compositions of planetary noble gases can be accounted for quantitatively by this mechanism, but it is necessary to assume an extremely high initial concentration of solar noble gases.

SUGGESTED FURTHER READING ON TOPICS IN CHAPTER 4

A number of excellent review papers on the isotopic dating of meteorites are available.

1 General: E. Anders, Meteorite Ages, *Rev. Mod. Phys.,* vol. 34, pp. 287–325, 1962, and also as chap. 13 of B. M. Middlehurst and G. P. Kuiper (eds.), "The Moon, Meteorites, and Comets," The University of Chicago Press, Chicago, 1963. The same author, in Origin, Age and Composition of Meteorites, *Space Sci. Rev.,* vol. 3, pp. 583–714, 1964, discusses all aspects of the meteorite problem, but of special interest is his interpretation of meteoritic gas-retention and cosmic-ray exposure ages in terms of collisions between asteroidal fragments. See also J. Zähringer, Isotope Chronology of Meteorites, *Ann. Rev. Astron. Astrophys.,* vol. 2, pp. 121–148, 1964.

2 Xenon (including Xe^{129}) in meteorites: J. H. Reynolds, Xenology, *J. Geophys. Research,* vol. 68, pp. 2939–2956, 1963.

3 Charged-particle tracks in meteorites, Pu^{244}: R. L. Fleischer, P. B. Price, R. M. Walker, and M. Maurette, Origins of Fossil Charged-particle Tracks in Meteorites, *J. Geophys. Research,* vol. 72, pp. 331–353, 1967.

4 Primordial noble gases in meteorites: R. O. Pepin and P. Signer, Primordial Rare Gases in Meteorites, *Science,* vol. 149, pp. 253–265, 1965.

5
PRIMORDIAL PLANETARY MATERIAL

> If the planets had a beginning, then either they must have formed from pieces of matter in an unconsolidated and chaotic state, which had been dispersed throughout a vast space before gravitational attraction gathered them into large masses; or else new planetary bodies have been formed from the fragments of much larger ones that were broken to pieces, either by some external impact or by an internal explosion . . . it seems likely that many of these original pieces would not have joined the larger accumulating planets, because they were too far from them or travelling at excessive velocities, but would have remained independent . . . continuing their journeys in space until each entered the sphere of attraction of some planet, whereupon it would fall, giving rise to the meteoritic phenomenon.
>
> E. F. F. Chladni (1794)*

We have seen that the meteorites are very old, the oldest samples of planetary matter presently available to us. Now it is time to consider the possibility alluded to in the Preface, that some meteorites may date back to the very beginning:

* From "Ueber den Ursprung der von Pallas gefunden und anderer ihr ähnlicher Eisenmassen, und über einige damit in Verbindung stehende Naturerscheinungen"; J. F. Hartknoch, publisher, Riga.

perhaps we have in our hands pieces of planet-rock that is still in the same state it first assumed when the planets were born.

What would we expect primitive planetary material to be like, from an astronomical point of view, if we had some of it? Before the solar system was born, its substance must have existed in a dispersed state, the volatile constituents (H_2, He, H_2O, CH_4) as tenuous gases and the nonvolatile constituents, such as the oxides of Si, Mg, and Fe, as small solid particles (dust). There seems little doubt that at some stage a clumping together or accretion of these condensed particles gave rise to the planets, so primordial planetary matter ought to consist of an aggregation of particles. There is no obvious way in which drastic fractionations of the chemical elements can have occurred during the accretion, at least so far as the very refractory and nonvolatile elements are concerned; therefore this primitive material should have a generalized and fairly uniform composition. Since planets and sun are thought to have formed from the same cloud of gas and dust (Chap. 6), the relative proportions of nonvolatile elements in both ought to be the same. Finally, since the various constituent dust particles may have formed in widely separated regions of space and time, and under very different circumstances, it would not be surprising to find them out of chemical equilibrium with one another.

These are the properties of chondritic meteorites, to a rather good approximation. Are the chondrites, or at least the unmetamorphosed ones, samples of primordial planetary matter, then? This is the key question that meteoric studies must answer. At the center of the question lie the chondrules, those curious spheroids of igneous rock that characterize the chondrites. Did preplanetary processes operate that can account for them, or must they have formed on or in the planets, in which case chondrites are secondary in origin, not primordial material? The answer is not yet known. In my opinion the evidence indicates that chondrites are indeed primordial planetary substance, but this point of view is disputed by several distinguished workers in the field.

H. C. Urey believes that the surfaces of the original planets melted soon after they formed, and that they separated (via igneous fractionation) into layers of diverse chemical composition. After these had solidified, some of the planets were reduced to rubble by mutual collisions. Subsequently this rubble tended to impact the surfaces of surviving larger planets, such as our moon. These hypervelocity collisions not only broke up the (previously fractionated) surface layers of the target planets, but in many cases melted the fragments. The droplets (chondrules) had compositions as diverse as the rock layers that gave rise to them; they froze quickly, then fell back to the planetary surfaces, later to become consolidated into chondritic rock. No truly primordial planetary material still survives.

Whether this hypothesis is more credible than the idea that chondrites are primordial material depends rather critically on the way Fig. 2-5 (the comparison of chondritic and solar compositions) is interpreted. What one sees in this figure seems to be a matter of personal philosophy. I have asserted in Chap. 2 that the agreement between sun and chondrites, while not perfect, is strikingly good. Professor Urey prefers to emphasize the discrepancies that appear in the plot as evidence that chondrites cannot have come directly out of the gas-dust cloud that predated the solar system.

Igneous processes have a strong tendency to chemically fractionate rocky materials. This has long been known from studies of terrestrial geology. Figure 2-6 shows how poorly the composition of the Earth's crust (a product of igneous fractionation) matches that of the sun, how it is enriched in some elements and depleted in others. Once fractionations of this magnitude had occurred on a planetary surface, what is the likelihood that pulverizing impacts or any other agency could remix the various igneous rock types so thoroughly as to restore the solar proportions of nonvolatile elements to the extent shown in Fig. 2-5? It seems small.

Several other objections can be raised to the details of Urey's hypothesis. Explosions caused by hypervelocity impacts on planetary surfaces would produce vastly more broken but unmelted debris than chondrules. Both theoretical considerations and experiments (subsurface nuclear weapons tests) indicate this. Thus chondrites ought to consist dominantly of angular fragments of various igneous rock types, of all sizes; but they don't. When chondrites do contain angular fragments, they are almost always of pre-existing chondritic rock. Fragments of normal igneous rock types are extremely rare.

Further, the process envisaged occurs on planetary surfaces, yet the cooling rates of most chondrites, as deduced from their metallic minerals (Chap. 3), indicate that they existed at considerable depth in their parent planets, some tens of km. It seems unlikely that such a great depth of impact debris could have accumulated. Hypervelocity impacts cause a net loss of material from small planets, i.e., more material is ejected at greater than escape velocity than the projectiles themselves contribute. For this reason a debris layer cannot grow to indefinite thickness, but should have a steady-state mean thickness not greater than the depths of the craters.

Urey's model depends upon magmatic fractionation to produce a diversity of rock and mineral compositions in planetary crusts; melted crystal fragments (chondrules) are then also diverse in composition, as observed. Yet recently a few chondrules of very peculiar mineralogy and bulk composition have been described (Fig. 5-1): magmatic fractionation in the Earth has never produced such compositions, to our knowledge. It appears that some other fractionating mechanism, outside our terrestrial experience, has operated on the chondrules.

Finally, we can now say with certainty that hypervelocity impacts on one particular planetary surface are not producing a material of chondritic character: the Surveyor 5 and 6 missions to the moon have shown the maria to be basaltic, not chondritic, in composition. Surveyor television photography also demonstrated that surficial impact debris on the moon consists almost wholly of fine-grained sand or dust, not chondrules.

Other writers have taken an intermediate view, that the chondrites are neither truly primitive nor highly evolved plan-

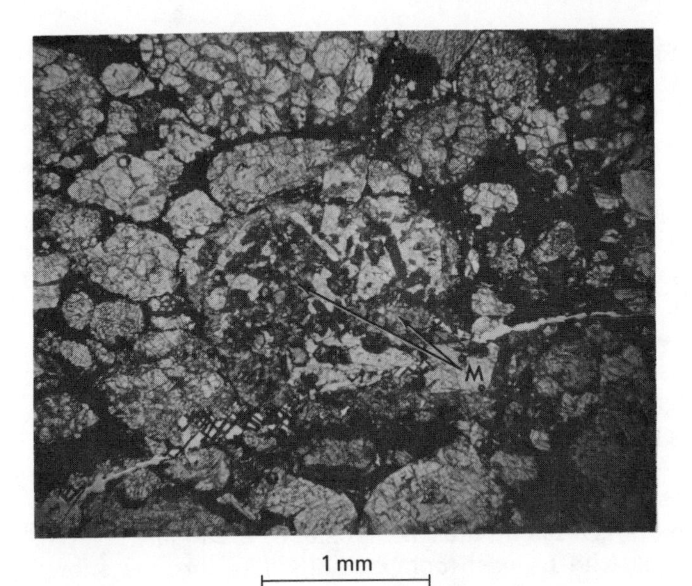

Fig. 5-1. Thin section photomicrograph of the Mezö-Madaras chondrite, showing a chondrule containing (M) the unusual bright-green mineral merrihueite, $(K,Na)_2(Fe,Mg)_5Si_{12}O_{30}$. This mineral has not been found on Earth, presumably because K,Na-bearing terrestrial igneous rocks are never so poor in Al as this chondrule. Surrounding chondrules are of more prosaic composition. *(Photograph courtesy of R. T. Dodd and W. R. van Schmus.)*

etary material, but something in between. A. E. Ringwood, of the Australian National University, and K. Fredriksson, of the Smithsonian Institution, believe that the primitive stuff of planets was finely particulate and undifferentiated, but that it contained no chondrules and no metallic nickel-iron. They propose that Type 1 carbonaceous chondrites (C1; Chap. 2) are surviving samples of the primordial material; Ni and Fe in these meteorites are present as oxides and sulfides. After bodies of this material had collected, high-energy events occurred that partly melted them, creating chondrules; and high-temperature carbon reduction produced metallic nickel-iron:

$$2(Fe,Ni)O + C \rightarrow Fe,Ni + CO_2 \text{ (gas)}$$

The high-energy events might have been hypervelocity impacts, which splashed up chondrule droplets just as Urey has proposed; or something akin to volcanoes may have been involved. Fredriksson has suggested that comets pelted the early planets and penetrated beneath their surfaces: there impact energy was transformed into heat, melting planetary material and vaporizing the ices of which comets are largely composed. The high-pressure gases so generated then blew showers of molten droplets out onto the planetary surfaces.

Several important objections can also be raised to these hypotheses. For one, the sudden melting and dispersal of C1 material, which is homogeneous on a rather fine scale, could not produce droplets that differed very much from one another in composition, yet we know from the varied mineralogies of chondrules that occur mixed together in any chondrite (Fig. 5-2) that their compositions are highly diverse. Chondrules with freakish compositions, like the one shown in Fig. 5-1, are especially hard to account for. The chondrule-making event must have been capable of chemically fractionating.

For another, C1 material contains a great deal of volatile material (organic compounds and chemically bound water): about 25% by weight. This would have to bubble out of the abruptly melted droplets of C1 material: certainly it is no longer present in the chondrules. However, the process can hardly have been 100% efficient; thus we ought to find that some of the chondrules still have a few gas bubbles entrained in them. But we don't: chondrules never contain bubbles.

It cannot be claimed that, by elimination, the writer has demonstrated that chondrites must be primordial material. There are difficulties with this concept, too, which will be discussed in Chap. 7. Yet in my opinion it comes closer to explaining the properties of chondrites than any of the other possible mechanisms that have been considered. The remainder of this volume will be devoted to a consideration of early events in the solar system and the origin of the planets, based on the premise that unmetamorphosed chondrites are samples of terrestrial planet materials as it first formed.

First it will be necessary to view the process of solar-

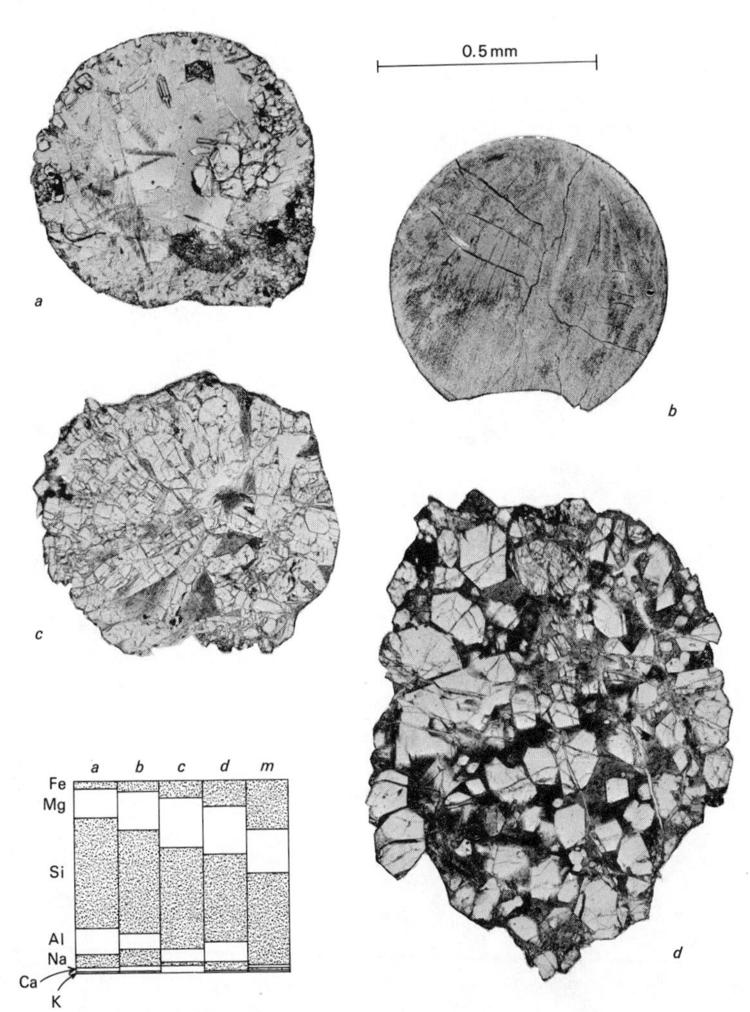

Fig. 5-2. Four of the types of chondrules that occur mixed in a single chondrite (Chainpur). (*a*) Glass with a few olivine crystals; (*b*) finely fibrous pyroxene and feldspar; (*c*) pyroxene crystals with glass between; (*d*) olivine crystals with glass between. The glass in (*c*) and (*d*) is partly devitrified. The graph at the bottom of the figure illustrates the diversity in overall composition (estimated) of these four chondrules in terms of atom percentages of major metallic elements. Column *m* shows composition of the merrihueite chondrule in Fig. 5-1.

system formation from a more fundamental perspective, that of the astrophysicists. This will be the subject matter of the next chapter.

SUGGESTED FURTHER READING ON TOPICS IN CHAPTER 5

Dissenting opinions as to the origin of chondrites:

1 H. C. Urey, A Review of Atomic Abundances in Chondrites and the Origin of Meteorites, *Rev. Geophys.,* vol. 2, pp. 1–34, 1964.

2 A. E. Ringwood, Origin of Chondrites, *Nature,* vol. 207, pp. 701–704, 1965.

3 K. Fredriksson, Chondrules and the Meteorite Parent Bodies, *Trans. N.Y. Acad. Sci.,* vol. 25, ser. II, pp. 756–769.

6 ORIGIN OF THE SOLAR SYSTEM

> I assume that all of the material of which the globes belonging to our solar system—all the planets and comets—consist, at the beginning of all things was decomposed into its primary elements, and filled the whole space of the universe in which the bodies formed out of it now revolve.
>
> Immanuel Kant (1755)*

Theories of how the planets were formed have been of two different kinds. One type of theory has depended upon extraordinary, and in many cases catastrophic, occurrences to produce the planets: a grazing collision or near miss by a passing star with the sun; fortuitous passage of the sun through a dense interstellar dust cloud; the explosion of a former companion star to the sun into an asymmetric supernova. The familiar Chamberlin-Moulton and Jeans-Jeffreys hypotheses belong in this category.

The other kind of theory has sought to explain the planets

* From "Allgemeine Naturgeschichte und Theorie des Himmels."

as normal by-products of star formation. Here the hypotheses of Descartes, Kant, and Laplace, and more recently Berlage, Alfvén, von Weizsäcker, and Kuiper, belong. All recent attempts to understand the origin of planets have been along this line. This is so partly because many of the properties of our planetary system follow as a natural result of the process of star formation as it is currently understood, but there also seems to be a philosophical reason for it: we have grown very humble about our position in the universe in recent years and find it intellectually distasteful to have to postulate that we and our insignificant Earth are the beneficiaries of any kind of special cosmic occurrence.

Kant's early concept, that a coming together of initially dispersed matter gave rise to the solar system, appears today to be essentially correct. Several lines of astronomical evidence indicate that stars are born inside vast clouds of gas and dust, light years in dimension, such as are visible in many parts of our galaxy (Fig. 6–1). Lyman Spitzer, of Princeton University, has played the most prominent role in developing this concept. The normal gas density between stars throughout the galaxy is ~0.1 H atom/cm^3, but in these clouds it can be as high as 1,000 H atoms/cm^3.

The interstellar gas is known from spectrographic studies to contain H, C, N, and O and small amounts of metallic elements. The dust grains are small, evidently in the micron size range (give or take a factor of ten). Their composition is not known; but because the grains are very cold (10 to 20°K), all kinds of atoms that collide with them except H, He, and Ne should stick and remain condensed. Thus their composition is probably generalized and their structure disorganized.

If stars are born in the interstellar medium, then the latter must have the same H- and He-rich elemental abundance pattern that we observe in the sun and stars. Nothing in our fragmentary knowledge of the interstellar medium contradicts this. (We have come to understand in large part why the elements exist in their particular pattern of abundances. Apparently the universe is far older than our sun, and generations of stars were born, matured, and died before the

Fig. 6-1. The Orion nebula, a bright, relatively dense cloud of gas and dust some 100 light years in dimension. The nebula contains numerous T Tauri stars and hot, young O and B stars, presumably just formed from its substance. *(Harvard College Observatory photograph.)*

events of this chapter occurred. The original stuff of the universe may have been entirely H, but thermonuclear reactions in hot stellar interiors are capable of combining H nuclei into heavier elements. The death of an old star sometimes comes in the form of a supernova, an explosion that mixes its substance, including the heavy elements it has created, back into the interstellar medium. Nuclear physics tells us which reactions would be favored in stellar interiors and the relative abundances of nuclides which they ought to produce; this abundance pattern turns out to be quite closely compatible with the observed solar and stellar abundances.)

Beginning with an abnormally dense concentration of interstellar matter—a cloud—we can define six stages in the course of events that we believe (mostly from theoretical considerations) leads to the formation of stars with planetary systems.

STAGE I—GRAVITATIONAL INSTABILITY

Interstellar clouds and the interstellar medium in general are almost everywhere in a state of approximate gravitational and pressure equilibrium. Movements of the interstellar material are mostly random, a roiling turbulence. But if a cloud region grows dense enough, and is large enough, instability can set in: the cloud's gravitational potential overwhelms other forces that act on the system, and it begins to pull itself together. Spitzer has estimated that a cloud whose mass is 10,000 to 20,000 times that of our sun will collapse self-gravitationally if its density exceeds $\sim$20 atoms/cm^3. A number of cloud complexes with roughly these properties have been identified in our galaxy.

STAGE II—COLLAPSE

As a cloud pulls itself together, gas pressure within it naturally increases, and gravitational potential energy is converted to other forms of energy, mostly heat. Much depends upon what happens to this heat: if it cannot accumulate in the cloud but is radiated away as fast as it is generated, so that the temperature in the cloud remains constant (*isothermal* collapse), then pressure will increase inversely as the volume decreases. On the other hand, if heat cannot escape (the *adiabatic* case), pressure increases inversely as volume raised to the $\sim$1.4 power. This makes quite a difference: collapse of a cloud to $^1/_{10}$ its original linear dimension (to $^1/_{1,000}$ its original volume) will increase pressure by a factor of 1,000 in the isothermal case but by a factor of over 16,000 if the system is adiabatic.

Gas pressure tends to oppose gravity, to hold a cloud apart; the pressure increase in an adiabatically collapsing cloud would soon halt collapse. However, since the tenuous clouds of interstellar gas and dust under consideration are quite transparent in the infrared, heat generated can be radiated away readily, and the system remains approximately isothermal. Under these circumstances, although the gradient of gas pressure increases as the cloud collapses, gravitational forces mount even more rapidly (as elements of the cloud draw closer to one another), so collapse continues unchecked. Cloud elements converge on the center at nearly free-fall velocity. The approximate time scale of collapse appears in Table 6–1.

Table 6-1

Stages in the collapse of an interstellar cloud (quantities very approximate)

Diameter of entire cloud, light years	Time required to reach next stage, years	Gas density, atoms/cm³	Minimum mass of "fragments" able to collapse independently, in solar masses
100	10^7	20	16,000
10	3×10^5	2×10^4	450
1	10^4	2×10^7	15
0.1	—	2×10^{10}	0.5

STAGE III—FRAGMENTATION

Since a cloud must be thousands of times as massive as our solar system in order to begin self-gravitational collapse, there must be some later stage at which it breaks up into subunits, each of which then evolves into a separate star. And, indeed, this is consistent with the properties of self-gravitating gas systems. The higher its gas density is, the less massive a cloud needs to be in order to separate itself from its surroundings and collapse independently. Density in the original interstellar cloud increases as the cloud collapses, so fragmentation into progressively smaller and more numerous masses becomes possible (last column of Table 6–1; Fig. 6–2). The new, smaller clouds or protostars would form around density fluctuations—regions of higher than average density—that existed in the original cloud. (However, the picture is complicated by the fact that magnetic fields would be embedded in interstellar gas clouds. A field of substantial strength would prevent fragmentation. It is necessary to assume either a very feeble initial magnetic field, or that the field and the gas ions it is attached to become separated from the cloud as it collapses.) Note that by the time the original cloud has collapsed to ~0.1 light year, fragmentation into systems as small as our sun will have become possible.

STAGE IV—ROTATING DISKS

The original interstellar cloud contemplated would have been rotating, very slowly, because the whole galaxy of which it is part is rotating (one revolution every 200 million years). It possessed angular momentum, which would have been conserved during its collapse. Because of this, the smaller the cloud (and later its fragments) shrank, the faster it (they) must have turned. The spin of the system had two important effects. First, it meant that the various cloud fragments did not fall straight to the center of mass of the system, as Fig. 6–2 implies, there to collide and coalesce again into a single supermassive star: instead of being straight, their paths were curved, and they fell in elliptical orbits around the center of mass of the system. Most of the protostars may have escaped collision under these circumstances.

Secondly, as the collapsing fragments turned faster and faster, they must have gone from roughly spherical shapes into increasingly flattened, spun-out disks. A. G. W. Cameron, of Yeshiva University, has concluded that some of the disks at this stage would have mass concentrated at their centers and others would not: they would simply be pancakes of gas and dust. Minor differences in density distribution within the cloud fragments, before they reached this stage, would have caused the difference. Cameron has suggested that disks of the first type ("axially condensed") give rise to stars with planetary systems, while binary star systems formed from the second type.

Fig. 6-2. Collapse, fragmentation, and subfragmentation of an interstellar cloud (sketched).

STAGE V—SLOW (HELMHOLTZ-KELVIN) CONTRACTION

Collapse of our disk was finally terminated when it changed over from an approximately isothermal process to an adiabatic one. This happened because, as its density increased, the substance of the disk became less and less transparent to radiant energy. At first the increase of opacity was due to a crowding together of the solid dust grains. Then, as the temperature rose, gaseous compounds such as H_2O, NH_3, and CH_4 evaporated from the dust grains and began to absorb radiant energy, converting it to rotational and vibrational molecular motion. At higher temperatures energy was absorbed by these compounds as they dissociated, and finally by the atoms themselves as they became ionized. So long as much of the released gravitational energy was spent in dissociating and ionizing, temperature did not rise too rapidly, and collapse continued: but as soon as there was nothing left to dissociate or ionize, the temperature (and therefore pressure) skyrocketed, and collapse ceased. The disk at this stage had about the same dimensions as our present solar system, according to Cameron, but was more massive, perhaps twice the present mass. Its center was at some tens of thousands of degrees Kelvin, and temperatures of thousands of degrees extended out beyond the present orbit of Jupiter.

The disk gradually separated into two discrete components: a relatively dense central condensation or *protosun,* surrounded by a thin, spun-out *nebula* of tenuous gas and dust. The protosun continued to convert gravitational potential energy to heat and continued to contract, but because now only its surface could radiate off heat, and the rest of the body could lose heat only by sending it to the surface relatively slowly via gaseous convection, the contraction was a slow and orderly process, not a collapse. In contrast to the previous collapse phase, the interior of the protosun was now in a state of hydrostatic equilibrium. Under these circumstances only about half the heat generated during contraction was convected to the surface and radiated off. Most of the rest was retained and progressively increased the internal temperature of the protosun.

Elements of gas in the nebula obeyed the same physical laws in orbiting about the protosun that planets do today as they circle the sun. The farther out an element of gas was in the nebula, the longer its period of revolution. Thus any particular element of gas was being passed by faster-moving gas interior to its position in the nebula, while slower-moving gas farther out in the nebula lagged behind. But in so doing, these gases would have dragged against one another: the gas element under consideration would have tended to retard the motion of gas immediately interior to it as it came past, and would itself have been accelerated by this interior gas. The magnitude of the drag, and whether or not it affected the nebula profoundly, would have depended on the effective viscosity of the gas. This in turn was determined by the density and degree of turbulence of the gas, and by the magnetic field strength in it. Cameron has postulated that the gas was highly turbulent, so substantial momentum transfer via viscous drag occurred. The net effect of drag between all the elements of gas that comprised the nebula would have been to slow gas rotation in the interior of the nebula and accelerate rotation at its outer edge. Slowed interior gases would have fallen into progressively smaller orbits—would have crowded in on the central protosun and joined it. Cameron believes that the protosun was quite small at first and grew by addition of nebular material to its present size. The accelerated rim of the nebula would have been spun farther out from the center.

STAGE VI—GENERATION OF NUCLEAR ENERGY

The protosun continued to contract slowly and its internal temperature to increase. Beyond a certain temperature the thermal motion of gas atoms is so rapid that they tend to undergo fusion (thermonuclear) reactions when they collide with one another. This stage was reached in the protosun when it had shrunk to roughly ten times its present radius and its central temperature had reached ~800,000°K. Further contraction and temperature increase brought into play a series of nuclear reactions, the net effect of which was to convert hydrogen to helium and to release vast amounts of energy. At last thermonuclear energy release became rapid enough to just balance the radiative heat loss from the sun's surface. Temperature in the solar interior was now, for the first time, held at values high enough to maintain gas pressures that were in turn high enough to balance the gravitational forces trying to pull the sun together. At this point contraction ceased, and the sun had evolved to essentially the state we observe it in now. Evolution since then has been subtle and very gradual. Calculations have indicated that the slow (Helmholtz-Kelvin) contraction lasted ~70 million years; i.e., this period elapsed between the end of free-fall collapse and the time when the sun took up its present quasi-stationary state.

By this time the nebula was probably gone. It appears that during the beginning of the Helmholtz-Kelvin contraction stage of star formation, much of the substance of the protostar is returned to interstellar space, by some type of violent solar-wind evaporation process. Nobody knows why this occurs: it is an observational fact, not a theoretical consideration. Members of one particular class of stars, the T Tauri variables (Fig. 6–3), are believed to represent the early contractive stage of evolution, and the spectra of these include absorption lines whose displacements to shorter wavelengths are interpreted to mean that the starlight has passed through gases that are moving away from the stars at velocities of 80 to 230 km/sec. The mean loss rate has been estimated at one Earth mass per thousand years. It seems highly probable that our own solar system passed through such a stage, and that the nebula could not have survived it. Large-scale mass ejections from a central condensation would sweep nebular gases away with them. The mass lost at the T Tauri stage might, indeed, consist entirely of nebular material. (It is to allow for mass losses at this stage that Cameron postulates that an axially condensed disk of ~2 solar masses gave rise to the solar system.)

At some stage and in some manner before the nebula was blown away, the nonvolatile elements in it condensed and agglomerated into solid bodies.* These had grown to substantial size (larger than dust) by the time the gas ejection began, else they would have been swept away with the nebular gases. In Chap. 5 it was argued that this primitive condensed matter was probably similar to the chondritic meteorites. In the next and final chapter the processes of

* Studies of stellar spectra have shown that the element lithium is much more abundant in young stars (T Tauri) than in mature, main-sequence stars. It appears that Li is destroyed in stars as they age, probably broken down into He by nuclear reactions in their hot interiors. In this light we can understand the Li discrepancy in Fig. 2–5. Apparently the chondrites still "remember" a time not far removed from our own sun's T Tauri stage when Li was more abundant in the solar system than it is in the sun now.

condensation and agglomeration and the growth of planets in the solar nebula will be examined.

Fig. 6-3. LkH_α 120, a T Tauri variable star (center). In this case the photograph has recorded a shell of bright nebulosity around the star, possibly ejected material, some 1,000 times farther from it than the outer limits of our own solar system. *(Photograph by G. H. Herbig, Lick Observatory; from L. V. Kuhi, Mass Loss from T Tauri Stars,* Astrophys. J., *vol. 140 pp. 1409–1433; copyright 1964 by the University of Chicago.)*

SUGGESTED FURTHER READING ON TOPICS IN CHAPTER 6

1 Historical review of theories of the origin of the solar system: R. Jastrow and A. G. W. Cameron (eds.), "Origin of the Solar System," pp. 1–37, Academic Press Inc., New York, 1963.

2 Star formation as it is currently understood: a brief discussion by L. Spitzer appears in "Origin of the Solar System" (above), pp. 39–53.

3 Origin of the elements: W. A. Fowler, The Origin of the Elements, *Proc. Nat. Acad. Sci.,* vol. 52, pp. 524–548, 1964.

4 Formation and evolution of the solar nebula: two papers by A. G. W. Cameron: The Formation of the Sun and Planets, *Icarus,* vol. 1, pp. 13–69, 1962; and Formation of the Solar Nebula, *Icarus,* vol. 1, pp. 339–342, 1963.

5 T Tauri stars: G. H. Herbig, The Properties and Problems of T Tauri Stars and Related Objects, *Advances Astron. Astrophys.,* vol. 1, pp. 47–103, 1962.

7
FORMATION OF THE PLANETS

> The ejected matter, at the outset, must have been in the free molecular state, since by the terms of the hypothesis it arose from a gaseous body; but the vast dispersion and the enormous surface exposed to radiation doubtless quickly reduced the more refractory portions to the liquid and solid state, attended by some degree of aggregation into small accretions. . . .
>
> T. C. Chamberlin (1904)*

We have a blurred picture of the primordial solar system, then, as a flattened disk of gas and dust, slowly revolving in the vastness of space. Its central region, where now the terrestrial planets orbit, was white-hot; here temperatures were so high that all the elements had vaporized; there was no dust. In the center of the disk, a tiny sun had taken form and was growing by drawing in gases from the surrounding nebula.

The central region would have radiated its heat away swiftly. Within months or a few years it should have cooled to temperatures at which the metallic elements condensed.

* From Fundamental Problems of Geology, *Carnegie Inst. Wash. Yr. Book No. 3,* pp. 195–254.

T. C. Chamberlin seems to have been the first to picture a primordial condensation yielding tiny particles or "planetesimals," although the cooling gas he had in mind came not from the interstellar medium but out of an already mature sun, plucked away by a passing star.

It is easy to see that in the outer reaches of the nebula, far from the warmth of the infant sun, condensation would have been most complete: even substances as volatile as the ices of water and ammonia would have condensed, if indeed in this region they had been evaporated from the original interstellar dust grains in the first place. In the warmer central nebula, only the more refractory metals and earthy compounds could condense. This must be why we have two classes of planets: the inner, terrestrial planets, small and of high density, composed of earthy materials; and the huge, low-density outer planets, composed largely of ices. When the condensate particles accreted into planets, the outer planets grew to greatest size because ice-forming compounds were much more abundant in the nebula than metals and earthy compounds, and because they collected material from larger volumes of the nebula.

FRAGMENTATION OF THE NEBULA?

Right at this point, however, a fundamental uncertainty about the behavior of the nebula looms up. Did it maintain itself as an unbroken disk until its gases were dispersed by a solar-wind-like mechanism; or did the pattern of fragmentation and self-gravitational collapse that led to the formation of the disk in the first place (as discussed in the last chapter) continue to operate, so that the nebula itself divided into numerous masses of collapsing gas, each in orbit about the infant sun? The question is partly answered by the Jeans instability criterion, a mathematical expression involving the density and temperature of a gas and the dimension of a hypothetical initial disturbance needed to give it a start toward fissioning. The Jeans criterion predicts that subfragmentation should indeed occur in gas systems as dense and massive as the postulated nebula.

Accordingly, G. P. Kuiper and later H. C. Urey have pictured an early solar system in which the nebula has broken up into a number of gas masses or protoplanets. In each, solid condensate particles fall to the center and collect into a terrestrial planet. Later the gas envelope is stripped away by the solar wind or some other agency. Urey believes that there were many such protoplanets, in which correspondingly small solid planets formed—typically of a dimension comparable to our moon. Most of these, the "primary objects," were later destroyed by mutual collisions, and a certain fraction of the fragments reaccreted to form the present array of planets, the "secondary objects." (Here and in the rest of this chapter the discussion concentrates on the inner, terrestrial planets.)

However, there are still important uncertainties in the concept of subfragmentation. The Jeans criterion does not take into account the gravitational effects of a massive body (protosun) in the gas system. Tidal disturbances due to the protosun would disrupt any protoplanets that formed, unless their densities exceeded a certain critical value (the Roche limit). Accordingly, in order to guarantee stable protoplanets, Urey has postulated a massive nebula, denser than the Roche limit. But this raises a new difficulty: D. Ter Haar, of Oxford University, and Cameron have pointed out that if the nebula were as substantial as that, then we would have tidal disturbances from the nebula itself, as well as from the protosun, to worry about. The problem of nebular subfragmentation has yet to be treated realistically. At this point we simply do not know whether the nebula fragmented or not.

The alternative is that the nebula remained intact, and that condensate particles in orbit about the young sun tended to clump together, ultimately building up into planets. The present chapter will be based on this concept, which seems more attractive to the writer.

What caused the particles to clump together? Gravitational forces are not enough. Gravity would become important only after a planet achieved substantial size: then it could draw particles in and hold them after they made contact. But in the beginning and early stages of agglomeration, it seems necessary to suppose that some gluey substance was present that caused particles to actively adhere whenever they touched. Urey first pointed out this requirement, at a time when he contemplated accretion in a nebula. Ice, organic material ("tar"), and electrostatic and ferromagnetic attractive forces have been suggested as "glues."

NATURE OF THE CONDENSATE

What was the nature of the solid material that condensed from the nebula? Chemical considerations tell us that nickel-iron metal, oxides, and sulfides and iron-magnesium silicates would have condensed directly from the vapor to the solid state. Under these circumstances, unless cooling had been very slow and regular (improbable in a turbulent gas), the product would have been tiny grains of dust, or possibly microscopic whiskers of mineral material.

With decreasing temperature (less than 500°C), any metal that had condensed would tend to oxidize and sulfidize, and silicates would tend to become hydrated (forming serpentine-like minerals). E. Anders has shown that a complex array of organic substances would appear—a potential "glue." Agglomeration of all these components would have yielded an earthy substance very similar to certain of the chondrites—the C1's (p. 44; Fig. 7–1).

If chondrites are truly primordial planetary material (Chap. 5) and the C1 and C2 chondrites have not even been metamorphosed (Chap. 3), then in these substances we can still study the original condensate particles from the nebula. The black matrix material between chondrules in C2 chondrites (Fig. 3–10) is quite similar to C1 material and can also be interpreted as a gluey agglomeration of condensed dust. But what of the chondrules themselves? Chondrules, or vestiges of them, are present in 98% of the chondrites—clearly they have played a key role in the condensation process, in at least one part of the solar system.

It is tempting to suppose that the chondrules, too, are particles of original condensate from the nebula. But the chondrules were hot liquid (igneous) droplets, as we saw in

Fig. 7-1. A piece of the C1 chondrite Orgueil about 1 cm in width. Chondrites of this type are complex mixtures of fine-grained, low-temperature minerals (chiefly magnetite and serpentine- or chlorite-like material) and organic compounds. They probably represent accretions of the first material (dust) that condensed in the solar system. *(Photograph from E. R. DuFresne and E. Anders, On the Chemical Evolution of the Carbonaceous Chondrites,* Geochim. Cosmochim. Acta, *vol. 26, pp. 1085–1114, 1962.*)

Chap. 2, and the physical requirements for condensation of silicate liquids from a gas of solar (=nebular) composition are prohibitive. Liquid condensation means high-temperature condensation; and at high temperatures relatively high vapor pressures of metallic elements are required before any material will condense. Total gas pressure in a nebula of solar composition would have to be some 20,000 times higher than the required metal vapor pressures, since H and He are that much more abundant than metals. It turns out that 100 atmospheres or more of gas pressure would be required to condense silicate liquids in the nebula (Fig. 7–2). Such pressure seems out of the question; the usual estimates of nebular pressures run to 0.0001 atm or less.

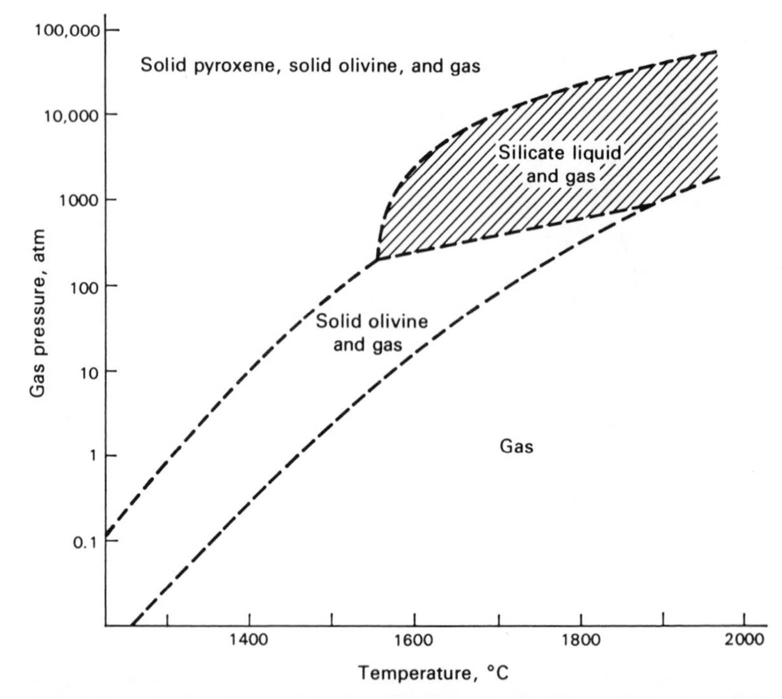

Fig. 7-2. Phase diagram for a mixture of H, O, Si, and Mg, in solar proportions. Only in the shaded field, at relatively high gas pressures, could liquid droplets (chondrules) be at equilibrium with a nebula of solar composition. The first condensate to appear, whether solid or liquid, is Si-poor (olivine-like); as the condensate cools (shifts leftward in the diagram), reaction with the gas phase increases its Si content.

THE FORMATION OF CHONDRULES

It appears more likely that already-condensed material (dust and C1 accretions) in the nebula was subjected to impulsive high-energy events of some kind, and transformed into chondrules. The variable luminosity of T Tauri stars suggests a violent and spasmodic activity at this stage, and possibly its effects are felt in a surrounding nebula. I have suggested that shock waves propagated through the nebula would momentarily compress and heat it. More recently,

F. L. Whipple, of the Smithsonian Astrophysical Observatory, and Cameron have pointed out that electrostatic charge separation may well have occurred in a nebula that was turbulent, dust-filled, and probably nonconducting (little ionized). If so, discharges—lightning bolts—would have occurred periodically, melting and vaporizing nearby solid particles. It is possible, in fact, that gas drawn into the discharge column by a magnetic "pinch effect" could sweep fine dust along with it, concentrating it into chondrules in the column.

It is interesting to try to interpret the mineralogy of chondrules in C2 chondrites in terms of remelting of nebular condensate before accretion. Olivine and pyroxene are the principal high-temperature minerals, both in chondrules and as smaller grains scattered in the matrix. There is considerable variation in the abundance of these minerals and the pattern of their Fe^{++} content among the C2's. Two extreme examples are shown in Fig. 7–3. Cold Bokkeveld is mostly matrix: chondrules are very rare; pyroxene and metallic nickel-iron are almost nonexistent. Renazzo is full of chon-

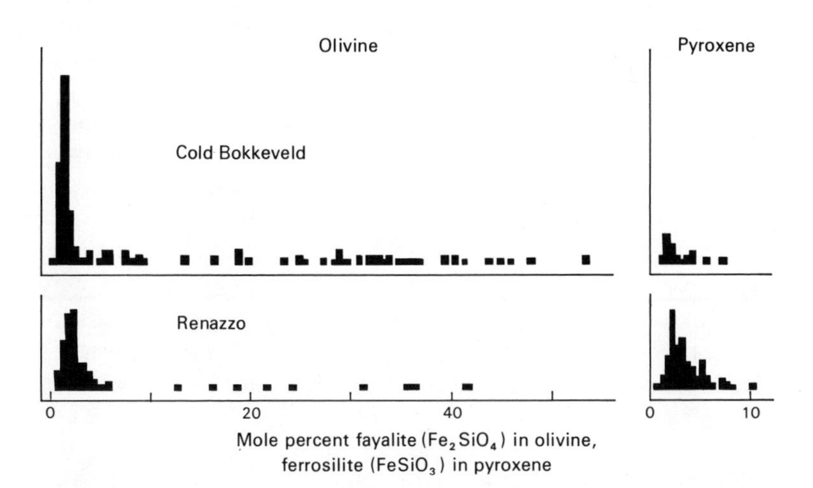

Fig. 7-3. Random surveys of mineral grains in two C2 chondrites, made with an electron microprobe. Histograms show relative abundances of olivine versus pyroxene, and Fe^{++} distribution (in terms of mole percent fayalite and ferrosilite) in each.

drules, and pyroxene is at least as abundant in them as olivine. Fe distribution in the pyroxene is similar to that shown for olivine. Metal is very abundant.

C2 matrix material has a composition quite close to that of an iron-rich (25 to 45 mole % fayalite) olivine. Melting of it would produce chondrules rich in olivine of this composition. Note that the Cold Bokkeveld histogram contains a cluster of olivines in just this compositional range. But what about the big spike of low-Fe^{++} olivines? It appears that something has acted here to reduce most of the olivine, removing Fe^{++} from the silicate melt-droplets by transforming it to metallic Fe. The H-rich nebular gas was a powerful reducing agent; could it have reacted with the droplets to produce this effect? There are two difficulties: (1) A gas of solar composition would have reduced the droplets to much lower Fe^{++} contents than we find in the Cold Bokkeveld or Renazzo chondrules. The higher the ratio H/O in a gas, the lower must be the Fe^{++} content of liquids in equilibrium with it; thermochemical calculations tell us that the chondrules with low-fayalite olivine in C2 chondrites would have been at equilibrium with a gas in which H/O was 50/1 or 100/1, but H/O in solar gas is about 1,000/1. (2) Melt-droplets cannot have remained fluid long enough and reacted thoroughly enough with a gas of solar composition to be reduced by it, without having been affected by the gas in a more drastic way: Fig. 7–2 shows that, at nebular pressures, they should have evaporated away into the gas.

Difficulty (1) suggests either that we are on the wrong track entirely in trying to understand the chondrules, or that they were created at points in the nebula where, for some reason, the composition was *not* identical to that of the present solar atmosphere: oxygen was more abundant by a factor of about 20 than it is in the sun.

But how can this be? It is difficult to imagine a physical process that could have fractionated nebular gases so radically. Turbulence keeps gases well mixed, and it is hard to sort the molecules out and send the various species in different directions. Fortunately, more than just gases are involved. Some of the oxygen in the nebula was incorporated in solid compounds—Fe, Mg, etc. oxide and silicate dust grains, as already noted. And in cooler parts of the nebula, practically all oxygen not present in this form would occur in solid particles of another kind—ice crystals (solid H_2O). Fractionations between solid particles and gases are easy to accomplish—almost impossible to avoid, in fact, in natural systems. Gravity would have drawn dust and ice crystals toward the median plane of the nebula and concentrated them there (Fig. 7–4). In addition, solids would have con-

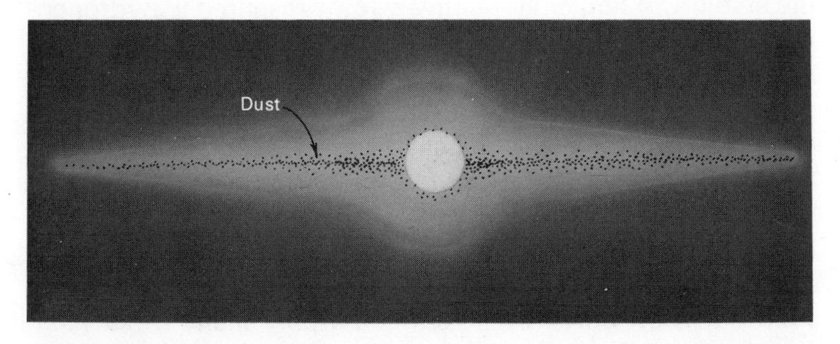

Fig. 7-4. Cross section of early nebula, showing dust concentrated near ecliptic plane.

centrated temporarily in "dead spaces" between the whorls and eddies of a turbulent gas. If metal oxides were the only solid particles to fractionate, they would have to concentrate locally to ~5,000 times their mean nebular abundance in order to raise O/H at these places to a value consistent with C2 chondrule compositions. If "snowflakes" were also involved, on the other hand, a concentration to only ~20 times the mean abundance would be required.

An oxygen-enriched environment for chondrule formation would also alleviate difficulty (2) above. Figure 7–2 holds only for a gas of solar composition; if higher concentrations of metallic elements and oxygen (relative to hydrogen) were present, then liquids would have been stable at pressures considerably lower than indicated in the figure: probably not as low as the nebular pressure of ~0.0001 atm mentioned earlier, but this may not have been the pressure operative when the chondrules were formed. The 0.0001 atm value refers to hydrostatic gas pressure in the nebula near the radius of the present asteroids, pressure due simply to self-gravitational forces in the nebula, which tend to draw the gases together. But the violent events postulated to account for the melting of the chondrules (shock waves or lightning) involve a momentary compression of the nebular gases to pressures temporarily much in excess of the hydrostatic value. When we take both effects into account—chondrule formation in momentarily compressed gas, and the stability of liquids at relatively low pressures in hydrogen-poor regions of the nebula—difficulty (2) begins to look much less serious.

Why would the process of chondrule formation yield products as dissimilar as Cold Bokkeveld and Renazzo? Possibly some batches of dust were affected more profoundly by the chondrule-forming events than others, before accretion. Suppose relatively few and feeble lightning strokes (or whatever) occurred in one particular region of the nebula. Some solid particles were vaporized, some only melted, but most remained unaffected. If any of the vaporized material was able to recondense as liquid droplets, these should have been at equilibrium with the surrounding gases; because of the reducing character of these gases, the droplets should have consisted of liquid iron and Fe^{++}-poor silicate liquids. Material that was only melted might not have had time to equilibrate with the nebular gases before it crystallized, in which case it would have remained rich in Fe^{++}, like the Cold Bokkeveld chondrules noted above that contain 25–45% fayalite olivine. The products—a few chondrules (some reduced, others not) and much unaffected dust—would have closely resembled the Cold Bokkeveld assemblage.

Where high-energy events were more frequent and powerful, vaporization-condensation might have predominated, yielding relatively more Fe^{++}-poor chondrules and liquid iron droplets (Renazzo). Generally higher temperatures in this situation could have left droplets in the liquid state for longer periods, allowing them to grow to larger size (via repeated collisions and coalescence) and react more thoroughly with their environment (increasing the Si content of the droplets as they cooled, so that in some cases pyroxene rather than olivine crystallized—see Fig. 7–2).

The metallic elements occurring in chondrules are volatile to different degrees. The fact that droplets tend to increase their Si content (relative to Mg) as they cool illustrates the greater effective volatility of Si oxide than Mg. Ca, Al, Na, and K also differ widely in volatility. Clearly the behavior of these elements would have been complex in the chondrule-forming situation described. The more volatile elements would have been distilled (lost selectively) from chondrules that were melted and only partly evaporated. Recondensing droplets would have incorporated greater or lesser amounts of these volatile elements depending on how rapidly they cooled and condensed. We can see why chondrules might be as diverse in composition as they are. And since distillation of silicate melts is a fractionation process quite outside our terrestrial experience, we cannot rule out the possibility that under rare and extreme circumstances it might yield chondrules of very unusual composition, like the one shown in Fig. 5–1.

OTHER TYPES OF CHONDRITES

The model discussed so far—condensation as solid dust particles, settling and concentration of the solids near the median plane of the nebula, and remelting and reduction of Fe by impulsive high-energy events—has been fitted to the unmetamorphosed C2 chondrites. All the other chondrites must have formed under similar circumstances, because they have similar bulk compositions and structures (chondrules). But there may have been differences. The chondrites may all have started out with Fe^{++}-poor chondrules, like the C2's, and then Fe^{++} diffused from matrix into chondrules during metamorphism, as was discussed in Chap. 3.* Or chon-

* Note that mixtures of chondrules (containing Fe in the metallic state) and matrix (containing oxidized Fe) in varying proportions would account for the differences in degree of oxidation among chondrites first noted by Nordenskjöld, p. 21.

drules in the ordinary chondrites may have contained substantial amounts of Fe^{++} from the beginning, reflecting formation under slightly different circumstances. Chondrules in the little-metamorphosed Type 3 chondrites have variable and often high Fe^{++} contents and may be examples of the starting material that was transformed by metamorphism into ordinary chondrites, as suggested by R. T. Dodd and W. R. van Schmus, of the Air Force Cambridge Research Laboratory.

Ordinary chondrites are poorer in various trace elements—Ag, Be, Br, Cd, Cl, Cs, Hg, I, In, Pb, Te, Tl, and Zn—than are C1 and C2 chondrites, which contain them in very nearly the expected solar proportions. The displaced In point in Fig. 2–5 demonstrates this fact dramatically. There is a correlation between the depletion of these elements and the degree of metamorphism observable in the various chondrites. Organic compounds and noble gases are also less abundant in metamorphosed chondrites. All these substances are relatively volatile and it is tempting to suppose that they were driven out of the parent chondrite planets, with varying degrees of thoroughness, by metamorphic heat. E. Anders has pointed out difficulties in this interpretation, however, and proposes that the varying abundances reflect different conditions of initial condensation and accretion. Specifically, the ordinary chondrites accreted at relatively high temperatures, at which the volatile substances in question had not condensed completely.

The highly reduced enstatite chondrites tell yet another tale. These may represent material condensed and remelted in unfractionated nebular gas, with O/H ~1/1,000. Or they may point to conditions of metamorphism that brought about Fe^{++} reduction in the body of the planet.

These considerations leave many questions unanswered. We wonder how, at some early stage, Fe was fractionated from the other metallic elements, producing the variability in Fe/Si between chondrite groups (Fig. 2–4) and between chondrites and the sun (Fe/Si appears to be about five times higher in the chondrites). Also, C2 matrix and C1 material are not identical in all respects to the vapor-solid condensate we would expect from nebular gas. Thermochemical considerations tell us that sulfur ought to have played a more important role. Obviously the picture of chondrite formation presented above is, at best, incomplete and oversimplified.

A SUMMING UP

Let us now gather the threads of this book and try to visualize the origin and evolution of a parent meteorite planet. We see, about 4.6×10^9 years ago, dust condensation from a primordial nebula; remelting of the dust by high-energy events, to form chondrules; accumulation, with the aid of a sticky binder, into asteroidal-sized planets; the dissipation of the nebula. A powerful, short-lived internal source of energy (probably Al^{26}) quickly heats the planet. Much of the chondritic rock is severely metamorphosed. Some is actually melted and separates into metal and silicate layers. The planet cools within 10^8 years and spends the rest of its life as a cold, inert body in space. It collides with other asteroids and is progressively reduced to rubble, which is dispersed in space. Some of this debris falls to Earth, for us to wonder at and puzzle over.

THE LARGER TERRESTRIAL PLANETS

Assuming that this picture is correct for one part of the solar system, presumably the asteroid belt, how applicable is it to the origin of the other terrestrial planets? We shall not know until primitive material has been sampled in other parts of the solar system. Certainly there were some differences. For example, Urey has stressed the fact that the terrestrial planets vary in density, those farther in being denser (Table 7–1). Apparently their content of Fe is higher. The process noted above that fractionated Fe among the chondritic meteorites must have had farther-reaching effects, through the whole system of inner planets.

P. Gast has presented evidence that the composition of the Earth is not identical to that of chondrites, the Earth being

Table 7-1

Densities of the planets (after H. C. Urey)

	Mass / Mass of Earth	Density	Uncompressed density*	Probable Fe content, %
Mercury	0.0543	5.59	5.2	~57
Venus	0.8137	5.12	4.0	~30
Earth	1	5.515	4.0	~30
Moon	0.0123	3.34	3.31	~10
Mars	0.1077	4.1	3.7	~26
(Chondrites)	—	—	3.65	~26

* The larger planets are dense partly because high pressure has compressed their interiors. In this column an attempt is made to estimate what the mean density of a planet's substance would be if it were all at zero pressure and 25°C.

depleted in relatively volatile elements such as Rb, K, and Pb. Possibly the planets accreted at progressively higher temperatures closer to the sun, so that by an extension of the mechanism proposed by Anders for ordinary chondrites, the more volatile elements were partially excluded from the Earth.

The moon presents a great problem. Since its density is much lower than that of the Earth (Table 7–1), it must be made of a fundamentally different material, having less Fe or more completely oxidized Fe or both. If it accreted at or near its present position, in orbit around the accreting Earth, why were the two products so different? Was there a fractionation during accretion that somehow sent mostly dense particles (for example, chondrules containing metal but poor in volatile elements because of their high-temperature origin) to the Earth and low-density dust (fully oxidized) to the moon? Or was the moon captured by the Earth, an intruder born in some other part of the solar system altogether? Dynamical arguments have been raised that support this alternative.

The newly formed meteorite planets were fiercely heated, probably by Al^{26}. There is no reason to suppose that the same early heat source was not present in the other terrestrial planets. Its effect would have been all the more profound in larger planets, from which heat escapes more slowly. It seems highly probable that the terrestrial planets were melted almost completely soon after their formation.

Our account of the formation of planets has consisted of a few facts, much conjecture, some questions, and a great many empty gaps. Meteorites have given us a wealth of information, but vastly more is needed. Most of all we need samples of primitive material from other parts of the solar system. Exciting times lie ahead when and if space missions return this kind of material. The Apollo mission to the moon provides the first prospect of doing so. It seems likely that the first returns will consist of volcanic rock, recording an early melting but not the actual formative process. But if subsequent missions carry out a more extensive survey, venturing into highland regions, they may conceivably find unmelted relics of the primordial moon. In the decades beyond Apollo there is the prospect of sampling the Martian surface, asteroids, and comets. Ultimately one can picture missions to Mercury and the satellites of the major planets. The generation of scientists that is privileged to study materials from these alien sources should be able to read, in some detail, the wondrous story of the creation.

SUGGESTED FURTHER READING ON TOPICS IN CHAPTER 7

1 The formation of planets: F. L. Whipple, The History of the Solar System, *Proc. Nat. Acad. Sci.,* vol. 52, pp. 565–594, 1964.

2 Gravitational instability in the nebula: G. P. Kuiper, On the Origin of the Solar System, *Proc. Nat. Acad. Sci.,* vol. 37, pp. 1–14, 1951.

3 Carbonaceous chondrites: B. Mason, The Carbonaceous Chondrites, *Space Sci. Rev.,* vol. 1, pp. 621–646, 1963; and E. Anders, On the Origin of Carbonaceous Chondrites, *Ann. N.Y. Acad. Sci.,* vol. 108, pp. 514–533, 1963.

4 Production of chondrules by "lightning": F. L. Whipple, Chondrules: Suggestion Concerning the Origin, *Science,* vol. 153, pp. 54–56, 1966; and A. G. W. Cameron, The Accumulation of Chondritic Material, *Earth Planetary Sci. Letters,* vol. 1, pp. 93–96, 1966.

5 Environment in which chondrules formed: J. A. Wood, Olivine and Pyroxene Compositions in Type II Carbonaceous Chondrites, *Geochim. Cosmochim. Acta,* vol. 31, pp. 2095–2108, 1967.

6 Underabundant trace elements in chondrites: J. W. Larimer and E. Anders, Chemical Fractionations in Meteorites. II. Abundance Patterns and Their Interpretation, *Geochim. Cosmochim. Acta,* vol. 31, pp. 1239–1270, 1967.

7 Formation of the moon: H. C. Urey, The Origin of the Moon and Its Relationship to the Origin of the Solar System, "The Moon," Z. Kopal and Z. Kadla (eds.), Academic Press Inc., New York, 1962; and E. L. Ruskol, On the Past History of the Earth-Moon System, *Icarus,* vol. 5, pp. 221–227, 1966.

INDEX